nanoculture

IMPLICATIONS OF THE NEW TECHNOSCIENCE

○○○

nanoculture

IMPLICATIONS OF THE NEW TECHNOSCIENCE

◇◇◇

EDITED BY N. KATHERINE HAYLES

GRAPHIC DESIGN BY DANIELLE FOUSHEE

◇◇◇

First Published in the UK in 2004 by
Intellect Books, PO Box 862, Bristol BS99 1DE, UK

First Published in the USA in 2004 by
Intellect Books, ISBS, 920 NE 58th Ave. Suite 300,
Portland, Oregon 97213-3786, USA

A catalogue record for this book is available from the British Library.

ISBN 1-84150-113-1

Designer: Danielle Foushée
Copy Editor: Holly Spradling

Printed and bound in Great Britain by The Cromwell Press, Wiltshire.

contents

Acknowledgements

This book has a dual purpose: it is intended to function both as an independent work of scholarship and as a complement to the nano exhibit mounted at the Los Angeles County Museum of Art from December 14, 2003 to September 1, 2004. Nano is being developed under the creative leadership of Victoria Vesna, internationally recognized media artist and chair of the Department of Design/Media Arts at University of California, Los Angeles, and James Gimzewski, a nanoscientist and member of the chemistry department at UCLA whose research has been recognized, among other honors, by the Feynman Medal. The nano exhibit provided focus and inspiration for this book; in addition, I thank Victoria and Jim for their support in publicizing and disseminating it, as well as for valuable feedback in responding to cover designs and graphic layout. Like *Nanoculture,* the *nano* exhibit is a work of many hands, and the more than thirty graduate students, faculty, and staff members whose work made the exhibit possible also contributed, through their work on the nano exhibit, to the context for *Nanoculture.* Bob Sain, Director of the LACMALab where nano is mounted, helped to keep us all sane along the way; Carol Eliel, LACMA curator of modern and contemporary art, and Kelly Carney, also of LACMA, provided encouragement and enthusiasm.

Roy Ascott, in addition to contributing the preface, has been instrumental in providing the guidance and support that has made it possible for the book to be published, and I am delighted that it is appearing in the series he edits. May Yao of Intellect Books gave invaluable advice and administrative guidance that helped the book move smoothly toward publication. Danielle Foushée enlivened the book with her graphic design, making it as vibrant visually as I hope it is intellectually. Michael Fadden provided valuable help with preparing the manuscript for publication. I am deeply grateful to UCLA for a sabbatical that gave me the time to work on the book, and especially to Thomas Wortham, chair of the UCLA English Department, for his generous support of my research in more ways than I can enumerate here. Nicholas Gessler contributed through innumerable conversations and ideas that have so woven themselves into the fabric of my thought that I no longer know which are his and which are mine. My greatest debt, of course, is to the contributors to this volume. With unfailing courtesy and promptness, they responded gallantly to my requests for revisions, information, and more information. Their insights, talent, research, and hard work have made this book what it is.

Preface

ROY ASCOTT

In seeking to promote creativity and scholarship in the culture arising from new developments in science and technology, *The Technoetic Arts Book Series* welcomes initiatives that employ a transdisciplinary approach to knowledge and informed speculation in pursuit of understanding. For this reason we are particularly glad that *Nanoculture: implications of the new technoscience* is the first book in the series to be published.

The transdisciplinary impulse exerts on its subjects both responsibilities towards specialisms as well as liberation from categorical constraints. When artists and scientists come together for more than mere dalliance, the *juissance* of their union (Barthes seems an apt reference here) depends on mutual understanding of each other's practice. In the case of the courtship of the arts and nanotechnology, go-betweens can also be enlisted to ease the union, in the form of literary speculation and scientifically informed fiction.

With exemplary clarity Katherine Hayles has edited a timely and insightful book, throbbing with ideas, and critically alive to the nuances of debate that the subject generates. Important too, is the transparency offered to the reader's understanding of the complexity of collaboration and transdisciplinarity involved in the creation of *nano,* the ground-breaking exhibition directed by Victoria Vesna and James Gimzewski, by which the book has been inspired.

Just over fifty years ago, an understanding of the fundamentals of matter (the *complexity* of structure in the material world) was sought in a similar bridging of art and science, which also resulted in a book and an exhibition. In that case, *Aspects of Form,* edited by Lancelot Law Whyte, accompanied *Growth and Form* exhibited at the ICA in London (1951). D'Arcy Wentworth Thompson stood then in relation to over-arching ideas of form and pattern, very much in the way that Richard Feynman is positioned, bottom-up as it were, in the debate around nanotechnology today.

The difference in editorial strategy between these two publications tells us much about the sea change in attitude that is now taking place intellectually and artistically in relation to science and technology. Where the earlier work saw the plastic arts as a *complement* to scientific inquiry, the present book sees the arts (now embracing fiction and literature as well as digital media) as *intrinsic* to the advancement of 21st century technoscience. Thus, it is a holistic attitude of mind and a transdisciplinary approach to knowledge that is seen as prerequisite to the construction of meaning, purpose and value in the emergent nanoculture.

Connecting the Quantum Dots: Nanotechscience and Culture

N. KATHERINE HAYLES, Department of English, UCLA

Imagine a world in which "utility fog" simulates a chair while you watch TV, mimics bathwater when you prepare for bed, and transforms itself into the bed you sleep on; a world in which micro-robots inside the body extend human life spans to centuries, manufactured bio-mechanical organisms clean the air and water, and material abundance is readily available to everyone on earth.[1] Such is the future envisioned by the proponents of nanotechnology. Poised between reality and dream, present and future, fact and fiction, nanotechnology has become a potent cultural signifier. Precisely because it is not yet clear if it will indeed be the "next big thing" or a blip on the screen, nanotechnology has attracted both skepticism and scientific research, along with a frenzy of entrepreneurial interest, government funding, and fictional speculation. Nanotechnology represents not so much a theoretical breakthrough as a concatenation of previously known theories, new instrumentation, discoveries of new phenomena at the nano-level, and synergistic overlaps between disciplines that appear to be converging into a new transdisciplinary research front.

"Nano" denotes one billionth of a meter, roughly the size of 10 hydrogen atoms; the DNA molecule, by comparison, is 2.3 nanometers in diameter. Nanotechnology is concerned with events and materials appropriate to this scale. Nanotechnology, in concert with nanoscience, for the first time in human history offers the ability to manipulate individual atoms and molecules, making possible radical new approaches to materials engineering. Consider, for example, work in 1985 by Konstantin Likhareva, a physics professor at Moscow State University, who along with his students Alexander Zorin and Dmitri Averin discovered they could control the movement of a single electron off a so-called "coulomb island," (a conductor weakly connected to the rest of a nanocircuit), leading to the possibility of a single-electron transistor, realized in 1987 by Gerald Dolan and Theodore Fulton of Bell Laboratories.[2] As silicon chip technology approaches the limits beyond which miniaturization is no longer feasible, discoveries like these promise to extend the miniaturization of information indefinitely using nanotechnology.

Also relevant is the work by Christopher Murray and his team at the IBM Thomas J. Watson Research Center using colloids for data storage (the colloids consist of magnetic nanoparticles in suspension, with each particle containing about 1,000 iron and platinum atoms). Spread on a surface, the nanoparticles crystallize into two- or three-dimensional lattice arrays. As George M. Whitesides and J. Christopher Love explain in "The Art of Building Small," "Initial studies indicate that these arrays can potentially store trillions of bits of data per square inch, giving them a capacity 10 to 100 times greater than that of present memory devices."[3] Research in other areas include "quantum dots" that fluoresce at selective wavelengths, making them useful tags to identify a variety of biological molecules; microfluidic devices that can deliver test solutions to specific parts of a cell under study; and dendrimers (branching molecules with large surface areas) that can transport DNA into cells for gene therapy or deliver drugs with precision directly to the organ or tissue that needs it.

Yet many of the visionary applications remain to be developed—or may prove impractical once development is attempted. According to Charles Ostman, the largest market share by far of present patents in nanotechnology is held in the cosmetic industry, particularly by Revlon, and involve new kinds of surface coatings that withstand water and have desirable hydration and other properties.[4] Better face creams are a long way from the global social and economic transformations envisioned by K. Eric Drexler, occasionally called "Mr. Nanotechnology" because of his influential 1986 book *Engines of Creation,* a visionary treatment of nanotechnology that popularized the research program laid out in 1959 by Richard Feynman's famous after-dinner speech (and later essay) entitled "There's Plenty of Room at the Bottom."[5]

Given that Drexler has founded the Foresight Institute to further research in nanotechnology, organized awards to recognize such research, published in 1992 a technical textbook on nanotechnology entitled *Nanosystems: Molecular Machinery, Manufacturing, and Computation,* co-authored with Chris Peterson and Gayle Pergamit in 1992 another book on nanotechnology, *Unbounding the Future: The Nanotechnology Revolution,* taught the first course ever on the subject at Stanford University, and tirelessly promoted nanotechnology in general, one might think he would have unquestioned stature in the field.[6] Yet many scientists look on him with suspicion and even disdain; one researcher working in the field told me a colleague was so upset with Drexler that, when seated with him at a conference, he challenged him to a fistfight. What evokes this kind of passion?

One answer, suggested by my colleagues Victoria Vesna and James Gimzewski in "The Nanomeme Syndrome: Blurring of Fact and Fiction in the Construction of a New Science," is that Drexler has it wrong when he proposes such mechanical devices as gears, pulleys, and conveyor belts made out of nanomaterials to fashion the replicators and assemblers necessary to turn out macroscale quantities of the new materials promised by nanotechnology.[7] These devices, they argue, are characteristic of the Industrial Revolution and are entirely retrograde when envisioned at the nanoscale. They think the inspiration should come from the realm of biology rather than mechanics. The issue is complicated by the ambiguous boundary between the biological and mechanical at the microscale (bearing in mind there is an important distinction between nanotechnology, which operates at the scale of a billionth of a meter, and cells, which are several hundred times larger). It is common practice to refer to the mechanism that maintains fluid equilibrium in a cell as a "sodium pump," which is a mechanical metaphor if not a mechanical actuality. The *e coli* bacterium incorporates a molecular motor that rotates a corkscrew tail functioning like a propeller (unlike the whiplike flagella of larger organisms). There are hundreds of other examples where biological processes are described in mechanical terms. R. Dean Astumian in "Making Molecules into Motors" describes research that constructed a ratchet and pawl from the organic molecules triptycene and helicene; other research mimicked the action of kinesin within a cell, which Astumian describes as a "molecular forklift" because it transports proteins along a nanoscale track called a microtubuline.[8]

As this example illustrates, part of the ambiguity results from a fuzzy boundary between literal description and metaphoric interpretation. Described as a forklift, kinesin sounds like a machine; described as part of the cell's interior, it sounds biological. Drexler himself frequently uses mechanical imagery when biological imagery would be as feasible. His definition of a machine—"any system, usually of rigid bodies, formed and connected to alter, transmit, and direct applied forces in a predetermined manner to accomplish a specific objective, such as the performance of useful work"—is sufficiently ambiguous as virtually to guarantee there can be no clear and stable boundary between the biological and mechanical.[9] One could almost say that the decision to locate micro-phenomena in the biological or mechanical realm is as much a function of the metaphors chosen as of the phenomena themselves. The choice of metaphor is consequential, for it lays down a linguistic track that thought tends to follow and suggests connections that bind new ideas into networks of existing conceptual structures.[10] What Drexler gains through his mechanical metaphors and imagery is the connotation that nanomaterials can be engineered, built, and controlled, as are mechanisms; what he loses is the connotation of dynamic change, mutation, and evolution characteristic of living matter. The subtext for his metaphoric choices centers on issues of control, for it is precisely the prospect that nanotechnology can replicate uncontrollably that is the greatest fear surrounding its development. By emphasizing the mechanical, he not only suggests that this technology can and will be constructed; he also minimizes the biologically-inflected implication that it may follow an agenda of its own independent of its creators' purposes.

Another—and perhaps more revealing—reason for the animosity toward Drexler permeates much of the scientific literature on nanotechnology. Typical is the headnote to the brief essay by Drexler in the *Scientific American* collection *Understanding Nanotechnology*. The editors acknowledge that "Many researchers in the field of nanotechnology discount the ideas of K. Eric Drexler and yet it is impossible to ignore his impact on the field."[11] The problem is spelled out more explicitly in the "Foreword" in a paragraph haunted by Drexler, whose presence is all the more notable because he is not mentioned by name. "But, if truth be told," the editors write, "—nanotechnology, apart from nature's realization, is still at present largely a *vision* for the future. But here vision and imagination are all important; in fact, there is ample room for scientific visionaries to coexist symbiotically with the futurists who fascinate, and sometimes prod, us with exotic dreams of worlds to come. Ultimately, however, all of our fanciful notions must be subjected to the refinement and distillation of true laboratory science. These hopes and expectations must be tempered by the reality that shortcuts generally do not exist; concrete bridges to new understanding must always emerge from a base of reproducible and verifiable scientific fact."[12] The problem with Drexler now becomes clear; scientists worry that he promises far too much with far too little experimental work to back up his claims. They fear he is squandering the cultural capital that science accumulates by patient and often laborious laboratory work, the source of "reproducible and verifiable scientific fact" for which no "shortcuts" can substitute. In a more sinister version of this objection, Drexler can be seen as garnering the glory of predicting a sweeping revolution for which others, with great cost and effort, struggle for years to accomplish even in small part.

The gap between vision and realization is apparent in Steven Ashley's "Nanobot Construction Crews," an analysis of the Zyvex Corporation. Zyvex was founded in 1997 by James R. Von Ehr II, a multimillionaire who deeply believed in Drexler's vision.[13] Von Ehr persuaded Ralph C. Merkle, a colleague of Drexler's in the Foresight Institute, famous for his work on cryptography for Internet security, to join it in 1999 (he has recently left Zyvex in July 2003 to take a position as director of the Georgia Tech Information Security Center[14]). Zyvex aims to create a nanoassembler, a nanoscale manufacturing device (assemblers are crucial in Drexler's vision of how nanotechnology can scale up to produce macroscale quantities of materials). To get to a nanoassembler, Zyvex starts from microelectromechanical systems (MEMS), structures that measure in microns and thus are 1,000 times larger than a nanometer. Created using lithographic patterns, the MEMS are fashioned with tiny "hands" that can then assemble much smaller components also made using lithography. Relying on precise positioning of the components, the manipulators can pick up and fasten the smaller units into assemblies using self-centering snap-connections. The other requirements to fulfill Drexler's vision are replicators, assemblers that can create copies of themselves, a capability still out of reach. The relatively slow progress (slow, that is, relative to Von Ehr's initial expectations) Zyvex has made illustrates the difficulties of actively realizing Drexler's vision.

According to their website, Zyvex now markets four products, three of them useful for nanoassembly: the S100 Nanomanipulator System, for use with a scanning electron microscope; the F100 Nanomanipulator system, used with a focused ion beam system; and Zyvex NanoSharp™ Probes, used to manipulate multi-wall and single-wall carbon nanotubes and nanoparticles.[15] At last report Zyvex was still far from profitability and also far from actually creating nanoassemblers. Writing in 2002, Ashley quotes Von Ehr as saying "This whole thing is a lot harder than it first seemed," acknowledging he has already spent $20 million on the project. Ashley also reports that "several scientists working in the field of nanotechnology derided Zyvex's scheme but requested anonymity to avoid protests from amateur nanotech enthusiasts."[16] Nevertheless, Zyvex's release of its four products was enthusiastically and uncritically announced by the NanoInvestor News website (October 4, 2003), evidence that the investor market remains hungry for nanotechnology.[17] A recent conference at California Institute of Technology for venture capitalists interested in nanotechnology cautioned that investors would be wise to think first about nanotechnology applications within established companies and industries (like Revlon) rather than to plunge into the untested waters of companies specializing directly in nanotechnology.[18] For better or worse, it seems that Drexler's vision remains a future hope rather than a present actuality.

This helps to explain why science fiction functions as the Other of nano-technoscience (Otherness implying a stigmatized partner that nevertheless remains essential for the originary term to constitute itself as such). At the same time that scientists welcome the visionary aspect of science fiction texts celebrating the technology's possibilities, they are also anxious to distance themselves from fiction, emphasizing as did the editors of the *Scientific American* collection that their work rests on "reproducible and verifiable scientific fact." It is remarkable how often science fiction is invoked in scientific and popular publications on nanotechnology, and just as remarkable how often it is positioned in opposition to what scientists actually do. Of course, science fiction texts dealing with nanotechnology are not always celebratory; frequently they imagine a dystopian future in which nanotechnology rampages out of control or is appropriated by an elite to oppress and control an underclass. Science fiction remains essential to nanotechnology precisely because it is not yet clear when and how the technology will become actualized. For the same reasons, nanotechnology continues to attract science fiction writers, who find in its nascent possibilities the potential for good storytelling, marvelous inventions that transform the world, and scary scenarios that fascinate even as they repel.

The *nano* Exhibit and Book Chapters

Like so much else in contemporary culture, this book had its origins in a coffee shop. On a perfect afternoon at Los Angeles Westside, I was meeting with James Gimzewski, Feynman Medal winner for his work with the scanning tunneling microscope (STM), and Victoria Vesna, internationally exhibited media artist and chair of the UCLA Design/Media Arts Department. Victoria and I had already collaborated on a number of projects, and she and Jim had collaborated on an interactive media art work, *Zero@Wavefunction*. Victoria explained that she had been invited by the Los Angeles County Museum of Art (LACMA) to create an interactive participatory exhibit in the Boone Gallery, a 10,000 square foot space that had formerly been a children's gallery and now was re-conceptualized as a "space for all ages," although its primary audience remained children. She suggested that the three of us take on the exhibit as an interdisciplinary project under the auspices of SINAPSE (Social Interactive Networks and Advanced Programmable Simulations and Environments), a non-center that Victoria and I, later joined by Jim, had initiated as a catalyst for interdisciplinary collaboration and dialogue among UCLA faculty and students. We had already planned the SINAPTIC Blowout, envisioned as a weekend event that would bring together graduate students from art, science and literature to brainstorm about interdisciplinary possibilities for collaborative projects. Thus began the work that would involve an interdisciplinary team, led by Victoria and Jim and including architects, museum staff, and forty UCLA faculty, graduate students, and staff. The work, culminating in the *nano* exhibit, is mounted at LACMA from December 14, 2003 to September 2004.

The exhibit takes as its throughline the idea of scale intrinsic to nanotechnology and nanoscience, creating playful interactions designed to give visitors experiences suggestive of what it would be like to be a nanoparticle subject to quantum forces, wave/particle dualities, and atomic and molecular interactions. The centerpiece of the exhibit is a large Inner Cell with a projected interactive floor showing a hexagonal grid that creates "gravity waves" when a visitor steps on it, low-frequency sounds, robotic spheres manipulated remotely by visitors in another exhibit module, and the *Zero@Wavelength* wall projection showing beach-ball-sized "buckyballs" (the carbon-60 soccerball structure discovered using nanotechniques) with which visitors can interact through their shadows. Also included in the exhibit is a "quantum tunnel" that plays with the scientific observation that electrons can "tunnel" through a barrier they do not have the energy to cross, the nanomandala, a projection of a sand mandala that is tied in with an exhibit at LACMA East created by Buddhist monks, the Fluid Bodies projection that has text dissapating, along with the image of a visitor's body, when a visitor walks in front of it, a studio area for painting and other activities, an interactive projection that takes gestures and transforms them into computer images (on loan from California Institute of Technology), and, running throughout the exhibit, text passages from scientists and science fiction writers speculating on the utopian possibilities and scary dangers of nanotechnology.

Nanoculture is intended to function both as an independent work of scholarship and a complement to the exhibit, exploring related issues through another mode of thinking and experience. Some of the issues it probes are explicit in the exhibit structure; others connect with the extensive discussions and debates that lie behind the exhibit's development but are not necessarily present in the exhibit itself; still others are suggested by thinking of nanotechnology as a cultural production as well as a technoscience. Intended as an art work, the exhibit also functioned as a case study in the possibilities and challenges of interdisciplinary collaboration, including the different ways in which artists, scientists, and humanists understand nanotechnology and the issues it raises. Different orientations and views were played out both in discussions and through the concrete work of making the exhibit, including how it would be constructed, what assumptions it would embody, what materials would be included or excluded, and what experiences would be crafted to what effect. Through the practical work of fabrication, participants came to understand better not only the different perspectives of the disciplines involved, but more fundamentally how disciplinary orientations manifested themselves in the obvious and subtle agreements and disagreements that emerged as the work progressed. The experience testified once again that interdisciplinary collaboration is not necessarily easy but can pay rich dividends in the quality of work produced, the insights gained, and the understandings broadened and deepened.

Two of the essays address the *nano* exhibit explicitly. In "The Invisible Imaginary: Museum Spaces, Hybrid Reality and Nanotechnology," Adriana de Souza e Silva presents a scenario of what a visitor sees as he interacts with the exhibit; she locates the exhibit in relation to changing views of how museum spaces should function. She traces a historical trajectory beginning with the modernist understanding of the museum as a "white cube" within which artifacts and objects are located, with the understanding that visitors see but do not touch or otherwise interact with the objects. The next phase is the advent of the virtual museum, initiated in the 1990s when various institutions experimented with the idea of a virtual museum as a supplement to the physical facility or as a stand-alone space accessed electronically through the World Wide Web. Contemporary developments have moved toward what she calls "hybrid spaces" in which physical locations merge with virtual functionalities. The *nano* exhibit exemplifies this hybridity, combining tessellated exhibit architecture with virtual projections, physical presence inside the Inner Cell with remote manipulation of robot spheres, and location within a museum gallery with an extensive website.

This hybridity is especially appropriate to the theme of the exhibit, for nano phenomena are far too small to experience directly. All knowledge of them is in some sense already virtual, mediated through precision scanning probe microscopes, data streams, and computer-generated visualizations. The exhibit's hybridity explores the significance of this split between the nanoworld and the "common sense" of macroscale intuition and experience, rendering the strange familiar and the estranging the familiar. For example, the visitor in the Inner Cell sees in the wall projection large beach-ball sized objects that seem very familiar. Yet when the visitor interacts with the projection using her shadow, she soon discovers that, unlike in the macroscale world, fast motions cause only small changes, whereas slow deliberate motions move the balls with maximum effect. The buckyballs also deform when "hit" by the shadow, following dynamics more akin to the deformation of atoms when probed by a scanning tunneling microscope tip than Newtonian laws of motion. At the same time, the exhibit subtly suggests through its tessellated architecture and text passages that the nanoworld is both an object of representation and the substance through which the representation is created. Ultimately all the materials in the exhibit, from the wall panels of the exhibit modules to the projectors mounted in the ceiling to the concrete floor, are made of atoms and molecules. The nanoworld is both on display (in its macroscopic representations) and invisible; touched, felt, and heard in simulated form and also actually present at every moment and in every cubic centimeter of space.

In "Working Boundaries on the Nano Exhibition," Carol Wald discusses the collaborative processes through which the exhibit was created. She decided early that she would write on the processes creating the exhibit and consequently attended meetings between architects, exhibit team and LACMA staff. At various points the collaborative process ran into difficulties, and her analysis astutely analyzes the sources of the problems as stemming from different understandings of what "collaboration" means. At issue also, she explains, was who would be in control of the "artistic vision" of the exhibit. Different stakeholders brought to the process different kinds of resources and expectations and divergent disciplinary perspectives. Negotiating issues of power, control, disciplinary orientations, resources, and the different institutional requirements of LACMA and UCLA was no simple matter. In the end, Wald suggests, it was everyone's commitment to creating the best possible exhibit that proved crucial.

The making of the exhibit as a collaborative process reflects some of the same interdisciplinary tensions and possibilities that inform the field of nanotechnology as a whole. The next group of chapters explores these by focusing on the connections between nanotechnology and science fiction. In "Nanotechnology in the Age of Posthuman Engineering: Science Fiction as Science," Colin Milburn cogently argues that nanotechnology's need to position itself as a legitimate science causes it to have an ambivalently double-edged relation to science fiction. Because nanotechnology is only beginning to develop, some of the most dramatic pronouncements of its future potential are either found in science fiction stories or, if put forward by a spokesperson like Eric Drexler, sound like science fiction even when meant as non-fictional extrapolation. Science fiction thus is essential to the field's articulation and, at the same time, that which must be excluded to establish the field's legitimacy. One of Milburn's striking examples is the famous after-dinner speech by Richard Feynman, "There's Plenty of Room at the Bottom." As Milburn demonstrates, this speech is widely regarded as the foundational moment of nanotechnology, referenced over and over again as the starting point of the field by writers who enhance the field's scientific credentials by connecting it to this world-famous scientist and Nobel Prize winner. Yet Feynman himself draws on Richard Heinlein's story "Waldo" (which he may have heard about from a colleague rather than read himself) for some of the best-known and often-cited examples in his speech. Revealing the penetration of nanoscience by science fiction even in its originary moment, this delicious irony indicates how inextricably bound up with the development of the technology are fictional extrapolations.

Also important in Milburn's chapter is his astute analysis of Eric Drexler's ambiguous relation to the field. As we have seen, Drexler is often excoriated for his excesses even as he is also acknowledged as having an undisputable impact. The controversy over Drexler also arose within the collaborative process that created the exhibit. As Carol Wald explains, the literature contingent wanted to include text passages from Drexler as well as from Michael Crichton's *Prey,* but Jim Gimzewski, Victoria Vesna, and some of the science students objected on the grounds that the science was not sound and the exhibit should not risk setting up these two figures as experts.[19] The literature people responded by saying that since Crichton's best-selling novel and Drexler's *Engines of Creation* have both been enormously important in influencing the popular perception of nanotechnology, they should be included because of their cultural importance, regardless of the soundness or unsoundness of the science. The debate mirrored the larger controversy over Drexler's position in the field: is the important thing the cultural perception of the new technoscience or the "reproducible and verifiable scientific fact" with which researchers must concern themselves? Given that nanotechnology remains largely a futuristic vision rather than an established technology and is moreover heavily dependent on funding from the National Science Foundation, the National Institute of Health, and various state legislatures as well as venture capitalists, it seems clear that *both* the cultural perception and the science are important. If the cultural perception turns largely negative, this could have a damaging effect on funding sources and consequently on the field's development; conversely, if the public responds to its utopian possibilities and positive transformative social impact, this would also be important. Researchers working in the field are of course acutely aware of both possibilities, which helps to explain their anxieties concerning Drexler.

In the specific context of the *nano* exhibit, the controversy over Drexler raised questions about how the exhibit was to be positioned and how the audience was to be conceived. Was the purpose to enact a vision of the future of the field that emphasized biological analogies and models rather than mechanical ones (Vesna and Gimzewski's view) and hence exclude Drexler because he has it wrong, or should it reflect the broader forces currently shaping cultural perceptions of the field? To keep peace in the family, the literature contingent conceded the point and agreed to minimize the presence of these two writers in the exhibit (although a small quotation from *Prey* is used and books by Drexler are included in the Resource Area). As the synergistic connections between Wald's and Milburn's chapters show, the controversies shaping the field as a whole also left their mark on the exhibit.

In "Less is More: Much Less is Much More: The Insistent Allure of Nanotechnology Narratives in Science Fiction," Brooks Landon explains not why nanotechnology needs science fiction but why science fiction needs nanotechnology. Noting the strong feedback loop between nanotechnology and science fiction to which Colin Milburn, Kate Marshall, and Brian Attebery also testify, he suggests that many science fiction writers basing their works on nanotechnology feel a sense of mission to help bring it into being. The connection between nanotechnology and SF is reinforced, he further suggests, because nanotechnology promises dramatic changes, and "change is the teleological heart of science fiction thinking." These changes are sometimes envisioned as First Contact, either through nano-artifacts left by aliens in our solar system, or as "aliens within" in which nanobots invade the body and, as in Greg Bear's *Blood Music,* refashion the human body along lines that make sense to them.[20] Other less drastic transformations are envisioned as arriving through the superior learning and education, as in Neal Stephenson's self-evolving, self-learning Primer in *The Diamond Age.*[21] Finally, nanotechnology promises to awaken the sense of awe and wonder that David E. Nye, among others, have called the Technological Sublime.[22] One version sees humanity transcending into a post-human future, sweeping into the dustbin of history the vulnerabilities of the "normal" human body. In his synthetic survey of science fiction works concerned with nanotechnology, Landon traces so many connections that one wonders how contemporary science fiction can be written *without* drawing on nanotechnology.

Positing nanotechnology as an enabling device for fiction writers, Landon's analysis tends to situate nanotechnology as a positive cultural force, notwithstanding his discussion of texts such as Michael Crichton's *Prey* that signal a clear and present danger in the runaway replication of nanotech devices.

He quotes Eric Drexler from *Engines of Creation* in a key passage where, describing runaway replication as "grey goo," Drexler acknowledges that "The grey goo threat makes one thing perfectly clear: we cannot afford certain kinds of accidents with replicating assemblers."[23] In the frequently cited article that Bill Joy wrote for *Wired*, "Why the Future Doesn't Need Us Anymore," Joy addresses this problem through the triple threat of nanotechnology, genetics, and robotics.[24] "It is most of all the power of destructive self-replication in genetics, nanotechnology, and robotics (GNR) that should give us pause. Self-replication is the modus operandi of genetic engineering, which uses the machinery of the cell to replicate its designs, and the prime danger underlying gray goo in nanotechnology. Stories of run-amok robots like the Borg, replicating or mutating to escape from the ethical constraints imposed on them by their creators, are well established in our science fiction books and movies. It is even possible that self-replication may be more fundamental than we thought, and hence harder—or even impossible—to control. . . . This is the first moment in the history of our planet when any species, by its own voluntary actions, has become a danger to itself—as well as to vast numbers of others." It is one thing to start the replication process, as Joy and Drexler acknowledge, and another to control it after replicators begin creating more of themselves, turning out more and more self-assemblies at exponentially increasing rates. Joy notes Drexler's prediction, for example, that artificial plants using nanotechnology to photo-synthesize will undoubtedly be more efficient than natural plants, creating the possibility that their reproduction could zoom out of control and kill off natural plants in a Darwinian survival of the fittest. Such a catastrophe would, of course, radically disrupt the food chain and cause disasters on a scale comparable to nuclear holocaust.

In "Future Present: Nanotechnology and the Scene of Risk," Kate Marshall uses the fact that nanotechnology lies largely in the future to connect it with the technology of risk assessment. She suggests that in a risk-conscious society, the present is determined by the future rather than the past. Further, risk-consciousness leads to action in the present to forestall future risk, sometimes leading to unexpected consequences requiring yet more interventions (a good example is the proposal made a few years ago to construct giant cannons that would blow bits of aluminum foil into the upper atmosphere to intervene in the greenhouse effect. If this bizarre proposal had been carried out, one can imagine that further technological interventions would have been necessary to cope with the effects of the reflective foil, and so on to infinity or global catastrophe). In the dizzying merry-go-round of anticipated risk leading to present intervention causing yet more risk and interventions, a reflexive loop forms by which society defines itself relative to the risk that it can neither completely avoid nor completely control. As Marshall puts it, a "risk society is sucked, face-forwards, towards the future's irresistible gravitational pull."

Marshall focuses on how this dynamic becomes enacted in representational forms. One of her examples, Houellebecq's *The Elementary Particles,* reveals as the narrative is concluding that it has been created by the posthuman products of a technological future as a "loving tribute to mankind." Fulfilling the nightmare scenario Bill Joy envisioned when he suggested that the human species has, through its own voluntary actions, become a potent danger to itself, the narrative's protagonist through his inventions uses technological evolution to evolve humankind out of existence. Thus the present-tense events of the narrative have already happened and led to the future that retrospectively relates them; since the supposed future has already happened, the seemingly open-ended possibilities of narrative complications are seen retrospectively as a teleological progression to an inevitable posthuman end. In the context of a risk society, the reflexive turns by which the present becomes the future becomes the past can be seen enacting a closure whereby our society is captured within the circular horizons of its own risk production.

This reflexive closure is closely related to the ambivalent transcendence that Brooks Landon identifies as centrally important in science fiction's fascination with nanotechnology. If the technology signals the end of humanity, one can perhaps mitigate the disaster by investing one's emotional capital in the posthuman descendents of humanity (as Hans Moravec does, for example, when he suggests in *Mind Children* that our true children and heirs are the intelligent robots that will become our evolutionary successors[25]). Alternatively, humanity itself can be seen as transcending its biological limitations, jumping the "singularity" (prophesized by Vernon Vinge, among others) that will utterly transform every aspect of human culture, society, and embodiment.[26] Brian Attebery uncovers another aspect to this dynamic in "Dust, Lust, and Other Messages from the Quantum Wonderland." Analyzing Greg Egan's short story "Dust" (the kernel for his novel *Permutation City)* and Geoff Ryman's *Child Garden* (among other fictions), he finds in them not so much futuristic scenarios as the recognition that in some sense we are already living in a nanoworld. Strange as its properties might seem from a macroscale perspective, it constitutes the stuff of which we are made. These fictions help us to realize, he suggests, that our experiences and perspectives are scale-dependent, holding true for only a small portion of the metrics by which we might understand the universe. Thus a premise that we take to be self-evident, "the belief that anyone can be isolated and self-contained," is, he concludes, "only sustainable within the narrow horizon of the human scale." Like Kate Marshall, the meaning he finds in fictions about nanotechnology lies not in prognostications about where we are headed but revelations about where we already are.

Whereas the previous essays concentrate on thematic connections between nanotechnology and science fiction, the next group exploring the relation of nanotechnology to literature focuses on the specificity of literature as literature—the materiality of poetic language, the literary enactment of narrative, and the literary evocation of wonder. Throughout the *nano* exhibit, links between the visualizations of nanoscience and their aesthetic and artistic implications are skillfully exploited. Because nano-phenomena can only be "seen" in highly mediated fashion, the technologies of visualization play a crucial role in the field's development. Data from scanning probe microscopes are routinely transformed into images. Although the shape of the images is dictated by the topography created by interactions between the probe's tip and atoms on the sample's surface (making the action of the tip more akin to feeling than seeing), important aesthetic choices also enter into these visualizations (for example, the colors used to represent different kinds of atoms). There are thus important areas of overlap between nanoscience and the visual arts. Indeed, as Carol Wald suggests in her essay, Victoria Vesna and James Gimzewski's common interest in visualizations may be part of the reason their collaboration has been so successful.

The case with literature is necessarily different. Although there are many instances where the visual play of the linguistic surface is important—in artists' books, concrete poetry, and line breaks in poetry in general—literature remains primarily a verbal art focusing (for print literature) on acts of writing and reading. To forge connections between nanoscience and literature, Nathan Brown opens a new line of inquiry by considering the scanning tunneling microscope as a writing instrument. He notes that one of the first breakthroughs in nanoscience to receive widespread press coverage was the image created by Donald Eigler and Erhard Schweizer of IBM Almaden Laboratory when, using the STM tip, they manipulated thirty-five xenon atoms to spell I-B-M. He uses this act of making—of writing at the limit of fabrication—as a metaphor for poetic engagements with the materialities of writing and language, putting both the technoscience and poetry "under the sign of building, as branches of materials research."

Of special interest in this regard is what Steve McCaffrey has called the "protosemantic" level of language, the processes through which the material properties of inscriptions and their combinatory possibilities become objects of attention before or along with semantic content. Working with texts ranging from Emily Dickinson's poem #640 ("I cannot live with You—") to McCaffrey's *Carnival*, Christian Bök's *Crystallography*, and Caroline Bergvall's *Goan Atom*, Brown shows how artistic engagement with the minutiae of patterns of ink on paper, visual topographies, and different combinations of letters to form clusters of evocatively related words function to create "quantum fluctuations" in significance and meaning.

Brown's discussion of Bergvall's *Goan Atom*, in particular, opens into a political critique of the nanotechnology industry and its participation in what he calls the "technocultural pornscape" of late capitalism. Amid speculation about nanotechnology's capacity to alter what Raymond Williams calls "structure of feeling," and amid utopian promises of global abundance in a nanocultural future, Brown re-engages the Heideggerian question concerning technology by asking what it might mean to "dwell poetically" at the limits of fabrication. While Brown responds to that question by gesturing toward ethico-political commitments beyond art and technics, he acknowledges that poetry may point us in the right direction by binding together inscription technologies, bodily enactments of reading and writing, and signifying practice, exploiting effects at different scale levels to create complex patterns that move us not only by what they mean but also how they mean. Both point beyond mere instrumentality—technology considered only as an instrument of capitalist investment, profit, and control—to interrogations of value, affirmations of complexity, and nuanced understandings of how the whole may be more than the sum of its parts.

Jessica Pressman's "Nano Narrative: A Parable from Electronic Literature" joins Colin Milburn and others in seeing narrative as essential to nanotechnology for a variety of reasons: to communicate its results to a wider audience; to enlist funding; to bridge the gap between the invisibility of the nanoworld and human understanding; and to envision the uncertain futures that the technology may bring into being. In fact, as Jerome Bruner has convincingly argued, narrative is an essential mode by which humans understand their worlds, and scientific inquiry is no exception. The work that narrative can do is especially important in instances where phenomena are inaccessible to direct sensory perception. Locating a common thread in techniques of mediation, Pressman points out that both nanotechnology and electronic literature depend on transcoding into binary code to be accessible to users. In the case of the scanning tunneling microscope, the analogue signals to and from the probe's tip are fed into a computer and there transcoded into binary code as a step to making the data available for interpretation; in the case of electronic literature, text, images, and sound are transcoded into binary code so as to be produced as a multimodal literary experience. Taking as her tutor text Erik

Loyer's Web-based literary work *Chroma,* Pressman suggests that knowledge of the nanoworld can illuminate the narrative techniques of *Chroma,* just as *Chroma* can suggest interpretations of nano-phenomena that give them human meaning and significance.

Pointing to the ways in which *Chroma* differs from print literature in its construction of the reader, Pressman shows that the user's responses are entrained and guided by the multimodal components of the work, including active animation (i.e., simulation), sound rhythms, flickering letters, mutating colors, and other non-verbal signifiers. Transformed from static artifact into computer-mediated process, the work is produced through interactions that, at a different scale and in different ways than the quantum wave/particle dualities of the nanoworld, are nevertheless similar to them in blurring the boundary between spatially discrete objects and distributed processes. Pressman intimates that if we can understand the ways in which apparently solid macroscale objects become processes at the nanoscale, we have an analogy that may help us to understand how literature transforms from the static and self-contained artifact of the book into the highly processual work of electronic literature.

Also important both to *Chroma* and nanotechnology is narrative. In its narrative, *Chroma* posits the existence of a "natural cyberspace" called Mnemonos, a space exterior to the human mind but nevertheless mind-like in that ideas are there apprehended directly, without the need of sensory perception. This unmediated encounter with ideas comes, however, at a price: when in Mnemonos, the researchers dubbed "marrow monkeys" are unable to communicate with each other or the world outside the Mnemenos. The challenge is to find a mode of representation that will make communication possible, which turns out to depend on narrative. Part of the meaning that the work goes in search of is the understanding that choosing a mode of representation, whether in literature or science, means buying into presuppositions that may not be explicit but are nevertheless crucial in determining how the results will be interpreted. In literature, these representational strategies include such well-known literary devices as character, plot, and narrative structure; they also include, Pressman insists, the medium in which the work is constructed. In nanoscience, some of the representational strategies are embedded in the instruments themselves; others are explicit choices about how to create visualizations and other interpretive devices. The service that a text like *Chroma* can render the popular understanding of nanoscience, Pressman suggests, is in reminding us that representations are never neutral, embodying in ways both obvious and subtle cultural implications that may be important in determining how the representations are understood and how they function in broader social contexts.

In "'What's the Buzz? Tell Me What's A-Happening': Wonder, Nanotechnology, and Alice's Adventures in Wonderland," Susan Lewak notes that the nanoworld is frequently compared to Wonderland because it operates according to very different rules than the macroscale world. To explore the roots of "wonder" as it is enacted in Lewis Carroll's canonical text, Lewak focuses on issues of scale and language. Like the "marrow monkeys" in Mnemonos, Alice discovers that in Wonderland language operates according to different rules; the poetry she had presumably learned to recite perfectly in the macro-world comes out all wrong in the miniature Wonderland. Drawing on Neils Bohr's understanding of the quantum world as a challenge to normative language based on macroscale experience, Lewak suggests that language and subjectivity are intimately related. Bohr argued that in the quantum world subject and object merge in what he called the "quantum of action."[27] Rather than existing independent of the observer, as objects do in the macroworld, entities in the quantum realm come into existence as measurable phenomena simultaneously with the methods used to observe them. "We are suspended in language," Bohr famously observed, a maxim that Alice verifies for herself when she finds out that her macroscale language apparently does not work in Wonderland as it did in the ordinary world.[28]

Moreover, since Alice's body constantly changes sizes in Wonderland, her sense of her own subjectivity is undermined. Lewak points out that Alice is able to take effective action and assert her agency when she begins to learn the different rules by which Wonderland operates. These rules are, Lewak emphasizes, not merely inversions of macroscale rules (as in *Through the Looking Glass)* but a kind of "no-sense" in which appearance and protocol are more important than substance and logic. As Alice masters these new rules, her sense of wonder fades. Wonder, Lewak suggests, is necessarily a transitory experience in which non-comprehension is an essential component. She speculates that as nanoscience and nanotechnology become more generally familiar, the wonder that presently accompanies them may diminish for the general public as the "buzz" migrates to another "next big thing." I suspect she is correct in thinking that the "buzz" requires constant novelty to maintain its "buzziness." As nanotechnology becomes an actuality rather than a futuristic vision, it will seem no more wonder-ful than, say, the computer sitting on my desk. Notwithstanding its ordinariness today, that computer is powerful enough to have stupefied John von Neumann, one of the inventors of the digital computer, if it could be transported back into the 1950s to arrive suddenly on his desk.

As these essays demonstrate, the connections of literature, art, and culture with nano-technoscience are as wide-ranging as they are subtle and complex. Narrative, language, subjectivity, visualization, desire, poetics, and the materiality of the signifier—all topics familiar to the arts and humanities—are also at issue in nanotechnology, either directly or through the cultural transformations it has the potential to enact. One of the controversial aspects of science studies is whether it has something to contribute not only to cultural studies, but to the sciences and scientific practices that are its objects of study. While opinion is divided on the issue, I believe that one of the goals of science studies, notwithstanding the problems attendant on advanced technologies, should be to repay in some small part the tremendous gifts that science and technology have given us. We can best achieve this, I think, through constructive interventions that enable scientists and technologists, as well as the society generally, to understand more fully the connections that enweb science and technology within the cultural and social processes that they affect even as they are affected by them. If *Nanoculture* achieves this goal even in modest measure, it will have earned its place in the world.

○○○

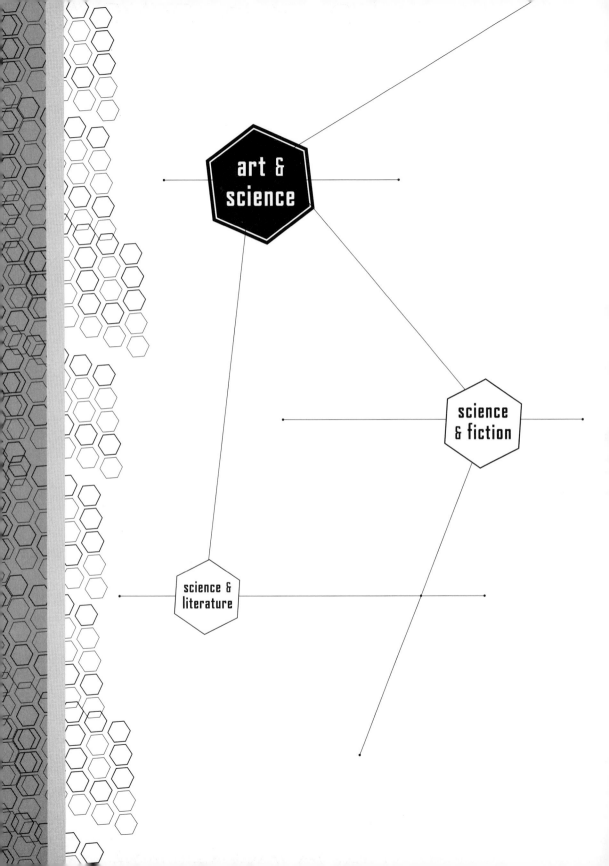

art &
science

science
& fiction

science &
literature

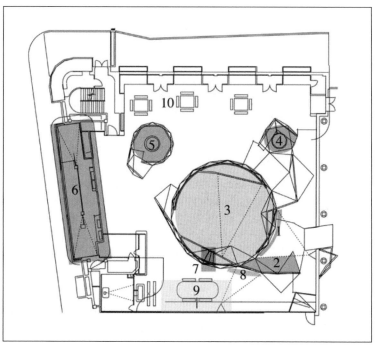

nano floorplan, Johnston Marklee & Associates

The Invisible Imaginary:
Museum Spaces, Hybrid Reality and Nanotechnology

ADRIANA de SOUZA e SILVA[1]

I. Prelude to the Imaginary: Experiencing *nano*[2]

Sketch of *ID Space*, by Ashok Sukumaran

Entrance[3]

As I walk into exhibition room, I see a "swarm" of cameras. One of them takes my picture, capturing my identity as soon as I enter *nano*. My face, as a product of this "unauthorized" surveillance, is then projected onto the wall in front of me, acquiring visibility among many other pictures, forming a huge hive-like structure. Each face inhabits a hexagon, which is also part of the graphite molecule structure. Standing in front of this construction, I realize that the hive-like projection is a database of several faces from people who have been in the exhibition before me. Observing the movement of people entering the exhibition room and being captured by the camera, I occasionally notice humorous aberrations, since the camera not only captures the visitors' faces but also everything it understands as a face, such as faces stamped on a visitor's T-shirt, for example. Having my face captured right in the beginning of the show reminds me about surveillance mechanisms. "But isn't this an exhibition about nanotechnology?" I ask myself. Why are there surveillance cameras at the entrance? How could I forget that one of the anticipated developments of nanotech is the ability to invisibly monitor and identify elements of one's identity? Nano, among other things, is about surveillance.[4]

The projection is the first movement interpolating the participant inside the nanospace. The animation of these "molecules" reminds me of the self-assembly mechanisms that many researchers expect will be used in nanotechnology. In the projection, each hexagon attempts to align with or attach itself to others, and groups combine into larger (and slower moving) aggregations. The hexagonal "molecules" with faces in them, resembling graphite, begin to shrink as new molecules are formed with other visitors' faces inside them. I feel as if my face is smaller than a protein. Nano is also about scale.

Sense Space

Now I am crossing a dark tunnel. I can barely
see, but rumbling sounds, echoing spoken words,
and the walls' texture transform the environment
into a different kind of sensory space, in which
I must use senses other than vision in order to
perceive. Suddenly my idea of what is stable and
physical is endangered, because I always thought
the most accurate way of perceiving the world
was by seeing. The dark atmosphere in the *Sense
Space* reminds me of many imaginary nightmares
commonly associated with nanotechnology, such

Sketch of *Sense Space*, by Ashok Sukumaran

as invisible nano-bots that invade the human body to destroy it, mind control through invisible
mechanisms, and molecular structures imperceptibly injected into the brain to manipulate people's
dreams. I wish to leave this space. Following the sounds and sensing the walls, I recall reading that
the nanoworld cannot be viewed, only sensed with the tip of a scanning tunneling microscope (STM)
that registers the topography of an atomic surface, an activity more like feeling than seeing. Then I
think about how it would be if I were one of the atoms being probed by the STM, as the *Sense Space*
flows into the *Inner Cell* of the exhibition.

Inner Cell

Sketches of *Inner Space*, by Ashok Sukumaran

Walking further through the tunneling *Sense Space*, I reach a circular cell. Cells are the cores of any
living organisms,[5] and a cell space is also the core of *nano*. The rumbling sounds I heard before are
coming from here. A projection can be viewed on the wall, and a different one on the floor; people
are walking through the environment; and four big spheres are rolling across the space. While I walk
slowly in order to sense the cell, I realize that I also affect the projected pattern on the floor. My steps
have the power to deform a glowing hexagonal grid, similar to a pattern of graphite molecules. The
deformation is a wave-like movement, transforming the static floor into a moving light pattern. The
reactive floor also triggers bass frequencies while I walk, mimicking wave behavior on a molecular
scale. It is as if I could hear the sound of an atom. Waves created by my movement over the floor
merge with waves produced by other visitors walking around, as well as by deformations induced by
the strange spheres. They are *robotic spheres,* automatically rolling over the floor apparently without
the need of human aid. Spherical shapes allude to atomic forms. Being able to touch these 3-foot tall
plastic spheres, rolling like giant atoms, makes me feel closer to the nanoworld. Then I think, for the

first time, that the environment around me and also myself are not only constructed by what I am able to see with my bare eyes. The glowing grid is also projected on top of the robotic "atoms," creating a three-dimensional curved surface. Everything is connected in this environment, like a propagating wave influencing all nearby elements.

Raising my head, I perceive a huge projection of cellular-like structures resembling buckminsterfullerene carbon-60 molecules, or buckyballs.[6] Participants' shadows are projected onto the same wall, sharing the virtual nanospace with the buckyballs. From time to time a new molecule grows on the wall, while some stand still. I realize that my shadow is able to move and deform these structures. However, not every movement affects the system. Abrupt and fast gestures are helpless.

Sketch of *Atomic Manipulation Space,* by Ashok Sukumaran

Atomic Manipulation Space

I exit the *Inner Cell* via another *Sense Space* and realize that these spaces function to connect the *Inner Cell* to the outside environment. I reach the *Atomic Manipulation Space,* which consists of a nine-sided table with a projection on top and four track balls on the edge. Getting closer, I perceive that the projection reproduces a bird's eye view of the same space I have been in before: the *robotic spheres* and visitors walking across the cell. Moving one of the track balls, I realize that they are interfaces used to control the robotic spheres in the *Inner Cell,* allowing me to be present in the former room, although not physically. If, as I thought earlier, the *robotic spheres* are atoms, my manipulation here resembles the manipulation of atoms through the STM.

Nanomandala

Exploring the space outside the cell, I find a dark room composed by a sand surface in the middle. Entering the space, I activate a projection over the sand, which images sand over a wide range of different scales, from visible sand grains to the invisible atomic structure. The transformation from sand to atom is inspired in the sand mandala created by Tibetan Buddhist monks for the "Circle

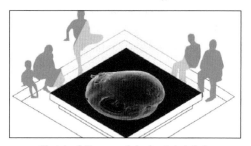

Sketch of *Nanomandala,* by Ashok Sukumaran

of Bliss" exhibit in LACMA East. I knew that the word mandala comes from Sanskrit and can be loosely translated as "circle," "whole," or "zero." A mandala can be regarded as a model for the organization structure of life itself, and there are many types of mandalas. The sand mandala exemplifies the impermanence of life and may take many days to be constructed. It is originally composed by colored sand made out of crushed semiprecious stones. These millions grains of sand are painstakingly placed on a flat platform and, after a period of days or weeks, are swept up into a jar and poured into a nearby water course to demonstrate the cycle of life. In dealing with atoms and cells, *nano* also deals with organization of living structures. The exhibition thus uses the *Nanomandala* to suggest a connection between two distinct processes of building the world from bottom up. While monks manipulate grains of sand as models for the organization structure of life, nanoscientists manipulate atoms, as the smaller known structures that construct the world.

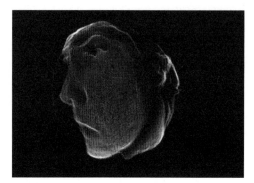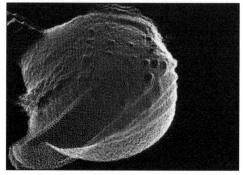

Quantum Tunneling, by Ashok Sukumaran and Osman Khan

Quantum Tunneling

Walking farther, I enter a dark environment with a mirrored floor. A camera stands on the top of a long and flexible metal structure. Grabbing the camera, I see that the static image projected on the wall in front of me starts moving. My image, converted into "particle clouds," begins to be graphically disturbed and altered. Looking back, I see that the movement of children running across a black tunnel has influenced my image. The mirrored floor over which they run reflects the actual the environment, creating a double sense of space. I don't exactly understand what is going on, but text on the wall outside explains that this part of the exhibit shows information being exchanged between two visitors standing at either end of the quantum tunnel, similar to electrons "tunneling" through an energy barrier because of quantum effects. I am curious about this phenomenon and decide to look for a more detailed explanation in the *Resource Room* of the exhibition.

Looking in from the Outside

Going towards the *Resource Room*, I find two holes on the outside wall of the *Inner Cell*, a lower one and a higher one. I am compelled to look, as the orifices radiate colorful bright lights, and discover a giant kaleidoscopic structure embedded among the wall panels. Looking through the higher one, I see the *Inner Cell* in a kaleidoscopic view. Besides the fracturing view of multiple perspectives, I hear narratives that

Sketch of *Kaleidoscopes,* by Ashok Sukumaran

sound science-fictional: "He's always wanted to become quantum dust, transcending his body mass. . . live outside the given limits in a chip, on a disk, as data, in whirl, in radiant spin, a consciousness saved from the void." (DeLillo 206)

Sketch of *Turbulent Text,* by Ashok Sukumaran

Text

A black-and-white particle cloud is projected on the outside wall of the *Inner Cell* adjacent to the kaleidoscopes. Walking in front of the projection I see that my movement across the space has the power to destabilize the particles. After the perturbation ceases, they reassemble into a phrase related to nanotechnology. Every time a visitor comes, he or she disturbs the text on the wall, making the particles rearrange into another different phrase. The dispersion of the image on the wall, like a swarm of particles, reminds me of Michael Crichton's passage in *Prey:* "A human body is actually a giant swarm. Or more precisely, it's a swarm of swarms, because each organ—blood, liver, kidneys—is a separate swarm. What we refer to as a body is really the combination of all these organ swarms." (Crichton 260)

At this point I am adjacent to the *Resource Room*, where books and other research material can be found on the tables. From this perspective, I look back on the exhibition space and notice the flowing lines of the architectural structures. I glance at a pedestal on which sits a leaded glass triangular model created by Buckminster Fuller and realize the same forms are used on the walls of the cells and modules. The model alludes to Fuller's Dymaxion Map, the only flat map showing the entire Earth surface without distorting the shape of the land areas and without splitting the continents. The idea of connection among parts, turning them into one and only structure, is present in the *nano* architecture, suggested by the flow from one space to another. Now I understand they are all simulations of a nanoworld where boundaries are fluid and solid objects melt with molecular motion. I start appreciating that everything around me is in fact made out of atoms, including my own body and brain.

2. *nano* and LacmaLab: changing the concept of museum spaces

Metaphorically injecting visitors into the invisible nanospace, *nano* challenges the traditional concept of what a museum is through three interconnected actions: enlarging what is supposed to be invisible; mixing virtual and physical spaces; and exploring the borderland between the real and the imaginary. These challenges, implied by the main exhibition pieces as well as by the exhibition space itself, are consistent with the main goals of LACMALab's Director, Robert Sain. LACMALab focuses on creating a new museum space that appeals to people of all ages, commissioning artists to create exhibits and construct participatory spaces.

nano is the fourth long-term exhibition developed by LACMALab, a research development unit of the Los Angeles County Museum of Art (LACMA). According to Robert Sain, "LACMALab is a new initiative designed to develop, test, and apply experimental approaches to engage the public—particularly children, teens, college students, parents, and seniors—with the museum's permanent collection and exhibitions."[7] The first show, *Made in California: NOW* opened in September 2000 and was up for ten months. Eleven California artists were commissioned to create interactive installations particularly to engage children and their families. The second exhibition, *Seeing,* lasted from November 2001 to September 2002. For this show, LACMALab commissioned nine Los Angeles-based artists with three broad challenging guidelines: works should explore the concept of "seeing;" they should incorporate at least one object from LACMA's permanent collection; and they should appeal equally to children and adults. *Making,* up from November 2002 to September 2003, was comprised of installations from five major art schools in Los Angeles: Art Center College of Design, California Institute of the Arts, Otis College of Art and Design, and the School of the Arts and Architecture at UCLA. Teams of student and faculty artists and designers were asked to create participatory pieces that investigate the process of making art.

For the first time, *nano* creates an overall concept for the whole space. The UCLA team, including media artists and nano-scientists led by media artist Victoria Vesna and nano-scientist Jim Gimzewski, together with writers led by N. Katherine Hayles involved in the production of the text passages in the exhibition space and this book, created the exhibit with the goal of producing a unified artwork that would suggest the participation of everything, including visitors, in the nanoworld, the space where the world's composition becomes apparent. In order to inject visitors inside the nanospace, the installations filling the exhibition deal with concepts such as scale, surveillance, boundaries, identity, seeing by feeling, and mapping invisible spaces. The interconnections between science, technology, art, and the humanities are expressed through architecture and installations that merge virtual and physical spaces, transforming the exhibition environment into a hybrid space.

LACMALab's initiative reflects a general tendency among the arts and museum fields towards inter-activity. It rethinks museum spaces in order to better integrate them with media arts (art mediated by technology). One approach is to transform traditional (impersonal, fixed, and neutral) museum spaces into participatory and interactive environments, thus promoting interaction between visitors and museum spaces, and among the visitors. In contrast to a conventional museum experience, it is no longer only the visitor who is changed by the space; the space is also modified by the visitor. *nano* is representative of initiatives by museums to incorporate media arts into their spaces, thus changing the relationship between museum and audience.

3. The construction of museum spaces

3.1. The physical museum

We can better appreciate the hybridization of contemporary museum spaces by looking back at how traditional modern museums were organized. Modern European and North American museums can trace their origins back to the 17th century, with the opening of The *Ashmolean Museum* in Oxford in 1683. The concept of a traditional museum developed as a place that encompassed a collection of artifacts of several types. These collections had their origins in medieval and Renaissance collections of wonders and rare objects, which belonged to private collectors and later were donated to public museums. According to Foucault, traditional museum spaces can be regarded as heterotopias, since they are spaces that juxtapose in a single physical place several virtual (non-present, but existent) places. Therefore the concept of the virtual as it is analyzed in the following pages is already contained in the idea of a museum. Foucault defines heterotopias as opposed to utopias: while the latter are sites with no physical location, heterotopias are physical places enacted by virtual components. In this sense, heterotopias call forth virtual intensities that are not yet actualized but are on the threshold of actualization. "Places of this kind are outside of all places, even though it may be possible to indicate their location in reality," argues Foucault.[8] Moreover, these places are absolutely different from the sites about which they reflect and speak. Similarly to libraries, which are collections of books from all places and times, museums are heterochronies or heterotopias of time. Consequently traditional museums, embedding within themselves their origins as a collection of objects from different times and places, include the seed of the virtual. Hybrid museums accentuate and develop this implication by positioning the virtual in a dynamic relationship to the actual rather than as something outside physical space.

Because their artifacts were supposed to be admired, museums developed as impersonal, neutral, and silent environments. The white cube was meant to create an isolated room disconnected from every aspect of outside physical space. The expectation was that museum visitors would be able to appreciate the artwork in a pure way, without any influence from outside reality. Reality was to be created by each individual object, which was in turn related to other virtual places. Traditional paintings and art objects were supposed to inhabit their own reality, which were not mixed with any other outside context.

To achieve this state, a certain distance between the viewer and the viewed object was required: no touch, no photograph, no loud talking. Walking through museum galleries, visitors created their own narratives that were not generally shared with other visitors. Moreover, the white cube was unchangeable. Granted that the perception of the room may change depending on what size paintings were hung on the wall and how art objects are placed in the environment, there was nevertheless no direct interaction between visitors and museum rooms. There was also no connection between visitors' movement and the shape of galleries, that is, the museum space was unaffected by the presence of visitors.

3.2. The virtual museum

The emergence of the World Wide Web in the 1990s facilitated the utopian vision of creating an ideal museum: one that someone could visit without being physically present. The virtual museum represented constant access to artworks from any point with an Internet connection. The traditional unchanging museum space contributed to the easy transference of museums from physical to digital spaces. As the physical space was not part of the exhibition, why not eliminate it? This thought led to the great misconception of the past decade, when digital spaces were in some instances regarded as replacements for physical environments. Digital cities were designed in order to create new types of sociability on the web, enabling users to make avatars and develop new social connections. As a result, websites were viewed as remote places that could be instantly accessed from whatever server in the world. The user was therefore no longer required to dislocate through physical space in order to reach remote locations and access information. Why go to a specific place if one can have everything via the Web?

Michel Serres exemplifies this argument pointing to the passage from definite places, such as the *Louvre* Museum on the Richelieu Street in Paris, to a rich place *(Riche lieu)*. This *'Richelieu,'* equivalent to the planet, is comparable to the net of all nets in which data and places virtually accumulate.

Because of the appearance of a global database that could potentially incorporate all data flows and information, digital spaces were frequently considered more important than physical places. "Why, then, a long street if it is enough to have a single place?" asks Serres (Serres 142). Virtual museums represented this possibility of an enormous database that could potentially contain all artworks.

According to William Mitchell, virtual museums have advantages over physical museums because "the exhibit material is kept on servers on a network, and viewers can be scattered at remote locations. It is not gallery capacity that matters, but server capability and network bandwidth" (Mitchell 59). Does accessing "information" about a museum replace the actual feeling of being inside a museum? Not really, for sure, but defenders of virtual museums also argued that they could offer far more choices for exploration than a large-scaled traditional museum. Although virtual museums would never mean the extinction of traditional museums, according to Mitchell, "as virtual museums develop, the role of actual museums will shift; they will increasingly be seen as places for going back to the originals." (Mitchell 60). As a result, one would see the work of art online, but one would go to a museum to see the original piece. Following the argument of Walter Benjamin's "The Work of Art in the Age of Mechanical Reproduction" (Benjamin 1990), the physical museum would be the place in which works would still have the aura of the original, and that is why they would remain significant.

If museums are there merely to display the original object and if many visitors did not care whether they saw the collection personally, changing viewing practices might arise that implicitly regarded an art object as representing only a specific amount of information. In this case, it would not matter which support was used to access the information: a web browser or a wood canvas. According to Claude Shannon's definition (Hayles 1999, 54), information is an immaterial entity that remains independent of the material substrate used to carry it. Considering that the concept of cyberspace has been based on the development of an information space, virtual museums have sometimes been viewed simply as information databases that can accumulate much more than the physical museum. In practice, however, a brief survey of museum websites shows that they are mostly constructed in order to support physical museums. They are useful to check a museum's opening hours, selected content of permanent collections and special exhibitions, but to date they have not begun to replace the traditional museum.

In contrast, Roberta Buiani points to a few Web initiatives that aim to create "real" virtual museums, that is, virtual places that have no original in the physical world and are not intended to supplement, or simulate, a traditional museum. Examples include the *Walker Art Center* section on net art, *Gallery 9,* and the *Uruguayan Museum El Pais.* While the first one focuses on net projects designed and conceived to be viewed solely online, the second example contains pictures of real painting and sculptures that belong to private collections. Both share the characteristic of displaying artworks that could not be contemplated by the common public in the physical world. Another singular example is the *Virtual Museum of Canada,* "which unifies under a single roof the resources of all Canadian museums" (Buiani 8). In this sense it could be defined as a heterotopia of a heterotopia, since traditional museums are already heterotopias. The *VMC* is a website about Canadian culture, and although much of its content can be in fact found in physical museums, there is no a single physical place in which the *VMC* would be contained. These last three examples differentiate substantially between virtual and physical museums, emphasizing that although they can be complementary, they may also have completely different purposes. There are also Web pieces created to be viewed online and, in this case, there is no point in showing them in a physical museum. By contrast, there is always a degradation of experience when seeing a traditional painting on the Web, and that is why websites of traditional art objects would never replace physical collections. Paintings acquire totally different meanings when looked at closely. The effect of light on canvas, the perception of brush strokes, and many other characteristics require the observer's physical presence for full appreciation.

Hence virtual museums can be divided in two types. One complements physical museums and is meant as a guide to their collections. The second type does not have a direct link with a physical building and provides the public with artworks that, because they are dispersed in different locations or are on-line creations, do not require or could not be exhibited in a single physical environment.

Even after the emergence of virtual museums, the function of museum spaces remained largely traditional. They still consisted of impersonal, neutral and quiet spaces. Nonetheless, questions about the museum's function and structure started to grow. The hypertext structure of the Web directed people's attention to more flexible ways of constructing narratives throughout the museum. Also, the emergence of online multiuser environments showed that an interactive and ever changing space could enhance communication among users. These developments catalyzed a new kind of approach. Could this ever-changing virtual space be brought into a physical 3-D environment?

3.3. The hybrid museum

The attempt to adapt museum spaces to show web art during the past decade was a challenge to most traditional museums. How to deal with projections and black boxes instead of white cubes? How to connect remote virtual environments to the actual space? With the development of media arts, new challenges were inevitable and many museums wondered how to adapt their spaces to deal with this new type of art. Unlike virtual museums, here the challenge arises from new interfaces used inside (or outside) the public space of a museum, in contrast to accessing a museum at a remote distance. Nomadic technologies and smaller interfaces, as well as real-time cameras and sensors, are being used by artists to convey their message in ways no longer compatible to the separation of the visitor from the exhibition space. Art pieces are popping out from the 2-D flat wall to inhabit a 3-D space that is, moreover, changed by the visitor's actions, making it unlike sculpture and the plastic arts. Also, this space is no longer disconnected from outside reality; rather, it brings the visitor (now called participant) into the artwork, creating new kinds of participatory spaces.

Considering that virtual museums did not replace traditional museums, it is probably true that media art museums will not do that either. Traditional museums will most likely remain established places constructed to show conventional art, such as paintings and sculptures. Nevertheless, the emergence of new forms of art that employ pervasive and mobile digital interfaces demand the creation of new types of museum spaces for those institutions interested in adding them to their collections. I shall call these new types of institutions hybrid museums. Hybrid museums have two main characteristics: (1) they merge the borders of physical and virtual spaces by means of the visitor's presence and mobility; (2) they promote direct interaction and communication among visitors and between visitors and the museum space.

According to Lev Manovich, "one trajectory which can be traced in the twentieth century art is from a two dimensional object placed on a wall towards the use of the whole 3-D space of a gallery".[9] However, he stresses that "this trajectory is not a linear development; rather, it consists of steps forward and steps back." The use of three-dimensional interactive installations dates back at least to the 1960s, especially in neo-concrete art. Brazilian artist Hélio Oiticica was among those who began to work with interactive installations, anticipating moving art off the canvas into the realm of life. By the mid-60s, Oiticica had abandoned traditional painting and sculpture in favor of freeform constructions *("Parangolés")*, such as capes meant to be worn or inhabited. Although his works did not employ electronic technology, they represented an important step toward reconfiguring the art object/viewer relationship. "Beginning in the 1970s", writes Manovich, "installation grows in importance to become in the 1980s the most common form of artistic practice of our times . . . Finally, the white cube becomes a cube—rather than just a collection of surfaces".[10] Accepting that information is never independent of the material interfaces that transmit it, we could argue that the present condition differs from previous concrete art and installations in the merging of virtual and physical spaces, a development hastened by nomadic interfaces and pervasive computing made possible by the increasing miniaturization of intelligent hardware.

Traditional museum practices, as well as artworks, have been affected by these new technologies. Many initiatives use nomadic and wearable interfaces in order to change the experience of walking through a museum. Flavia Sparacino's *Museum Wearable* (MIT) consists of a wearable computer that functions as a museum interactive guide. The *Museum Wearable* is an interface carried by the visitor, composed of an audio system and a "private eye" that personalizes the museum visit. For example, if the visitor spends a long time in front of a specific painting, the system delivers audiovisual information about that work. Therefore the guide is configured differently for each visitor. According to Sparacino, the goal of the project is to create "a system which can be personalized to be able to dynamically create and update paths through a large database of content and deliver to the user in real time during the visit all the information he/she desires" (Sparacino 2). This project follows the tendency of major Internet content during the 1990s to create private spaces inside public spaces. Websites such as amazon.com created personalized information for different types of users. In addition, chat rooms and personal messaging further increased the sense of personal spaces inside the mega spaces of the Web. Although these technologies changed the museum space to some extent, they did not transform the fundamental meaning of what a museum could be.

In contrast, new media arts, transforming neutral spaces into participatory environments, changed the meaning of the space itself from a silent place to an actively experiential communication and learning environment. There are basically two ways in which media arts contribute to the creation of hybrid (vivid and participatory) places. One occurs inside museum spaces; the other happens in public spaces. The exhibition *nano,* about which we have heard and which will be analyzed later in this chapter, exemplifies the first case of transforming an interior space. *nano* constructs an architecture that is as part of the exhibit as the installations it contains. The second case is represented by large-scaled interactive installations placed in public spaces and new media exhibitions in city spaces, which use existing architecture contained in public spaces and then modify their original meaning.

Representative of this second case is the work of Mexican-Canadian media artist Raphael Lozano-Hemmer. He created public installations that changed how people behaved and perceived public spaces. *Vectorial Elevation (Relational Architecture 4)* consisted of 18 robotic searchlights installed on the top of buildings around the Mexico City's Zócalo Square (in its first version). The lights were controlled remotely via a website on the Internet, allowing users to draw different patterns on a 3-D model online that were then displayed in physical space. The online participant could view the physical result of her drawing with the aid of three webcams placed on the square. This project linked digital space with physical environment in a non-traditional way. Generally, cyberspace is viewed as a place in which users can enter and create new digital worlds, which are not contained in physical spaces. With *Vectorial Elevation,* however, digital space was used to modify a physical plaza. Equally important was the way Mexican citizens reacted to the piece. The Zócalo Plaza was already an important reference point in the city. *Vectorial Elevation* increased its physical impact, since many people went to the plaza at night to observe the light patterns, transforming it in a vivid public space. The same reaction can be observed with other pieces by Lozano-Hemmer, such as *Body Movies (Relational Architecture 6)* and the recent *Amodal Suspension (Relational Architecture 8). Amodal Suspension* uses searchlights in a public space similar to *Vectorial Elevation.* However, instead of displaying online drawings, the searchlights "catch" still unread SMS (short message service) sent among users via the website or their cell phones. The messages are then encoded into a sequence of flashes and displayed in the sky around the Yamaguchi Center for Arts and Media (YCAM) in Japan. *Amodal Suspension* is virtual in a double sense: first because it connects users who are not physically present; second because it displays text messages before their actualization. It therefore catches virtual messages in a potential state, before they have been read, and transforms them into a physical entity.

Another example is The Chaos Computer Club's *Blinkenlights*, an eight-story façade used to play Pong or display large-scaled love letters, using cell phones as an interface. The project attracted a great number of people to Alexanderplatz in Berlin (2001/2002). Given these results, we can suggest that the importance of these projects is transforming circulation spaces, where people pass through but do not stop, into public places in which to gather. The space is no longer used for transit only but becomes a place where communication occurs and pleasurable experiences happen.

These works can be understood as attempts to transform spaces into places. Manuel Castells (Castells 409) creates a dialectical opposition between the *space of flows* and the *space of places*. The last one corresponds to the spatial organization of our common experience defined by cities and urban spaces, while the first one is a concept created to label a new logic of space structured on networks and flows of information. Castells suggests that the *space of flows* in traditional urban spaces transforms the city from forms into processes. Therefore, mega-cities no longer happen in a place but rather are "discontinuous constellations of spatial fragments, functional pieces, and social segments" (Castells 436). Felix Stalder, following Castells, observes that "the space of flows is not so much organized to move things from one place to another, but to keep them moving around."[11]

If we consider metropolises like Los Angeles as representative of circulation spaces, where people generally do not walk on the streets and prefer cars and freeways to move around, it is possible to argue that urban public spaces have become increasingly non-places. Although Castells argues that the *space of flows* is not a placeless space, places have progressively lost their importance in comparison to flows. Especially after the advent of advanced transportation technologies in the 19th century, people started to circulate faster across urban spaces, losing the capacity to communicate and interact to each other while in transit.

With the advent of the Internet, communication places partially migrated into digital spaces. Online multiuser environments, for example, can be perceived as places in which people talk and interact to each other, even if they do not share the same contiguous physical space. In the past decade, they have often been regarded as ideal communication places, and some commentators predicted they would replace public spaces in the role of promoting interaction among people. Nowadays, however, it is possible to observe a tendency to bring these gathering and communication places again into physical space. As we have seen, many artistic initiatives strive to transform public urban spaces by making people stop while in transit across the city. Such installation pieces have been successful in making the circulatory *space of flows* again into a *space of places*. To this extent, people no longer use urban spaces only to circulate and go from place to place, but rather start enjoying going to public places as their destination.

Related initiatives go beyond changing a single physical infrastructure or a single installation in public space. Huge media arts projects may construct an entire art space in order to "revive" city spaces and improve communication among people. These art spaces can be considered new types of museums, designed specifically to accommodate non-traditional art that deal with new digital interfaces. For instance, in the Ruhr area in Germany, old inactive coalmines are being transformed into exhibition spaces for media artists. The exhibition *Connected Cities* (1999)[12] transformed the area into a temporary laboratory in which artists represented the urban industrial situation as an enormous collection of cities connected not only through the usual transportation systems but also increasingly linked through invisible lines of communication such as networks and digital media. In this case, as well as in the previous examples, we can observe how embedding virtual into physical spaces can change our relationship to public spaces.

In this context, the concept of the virtual as it relates to a museum no longer means merely a website that can be accessed remotely. A different sense of the virtual is constructed as a potentiality that can be actualized. In a museum context, these virtualities are actualized by visitor's actions in physical spaces. Virtualities are potentialities always ready to emerge and to reconfigure the reality in which they appear.

4. The construction of hybrid spaces

4.1. Re-creating reality as an emergent potentiality: the virtual

Frequently the virtual has been considered to be opposed to the physical, mainly because cyberspace has often been considered an immaterial space. A hybrid space occurs when one no longer needs to go out of physical space to get in touch with virtual (or potential) realities. Hybrid spaces have three main characteristics: (1) the merging of borders between physical and virtual spaces, (2) the use of nomadic and pervasive technologies as interfaces, and (3) mobility and communication in public spaces. Hybrid spaces fold the virtual as potential into the nearby physical space, blurring the borderlands where the virtual transforms into the actual, and the actual fades back into the virtual.

There is a dynamic interplay whereby virtual becomes actualized and the actual becomes once again virtual. In this sense, hybrid space is different from an augmented reality that superimposes graphic or sound information onto a view of the real world, or an augmented virtuality that refers to "augmenting or enhancing the virtual world produced by a computer with data from the real world." (Ohta and Tamura 2). Hybrid spaces are also unlike mixed reality as described by Paul Milgran (Milgran in Ohta and Tamura 10). Milgran suggests that a mixed reality occurs when "it is not obvious whether the primary environment is real or virtual," creating a RV (real-virtual) *continuum*. Hybrid reality, by contrast, does not oppose real and virtual; it includes the virtual in the scope of the real. Both the virtual and the actual are real; the difference between them is not reality or unreality, but rather their positions in the temporal dimension.

In contrast to this view, virtual spaces have usually been related to static interfaces used to connect to the Internet and Web, such as desktop computers, large monitors, and corded mice. One needed to "enter" the Internet in order to "inhabit" a virtual space, by implication temporarily leaving physical space behind. Virtual reality has also often been connected to a Platonic logic of representation, later developed by Jean Baudrillard (Baudrillard 1-48), as a simulated (hyperreal) space that could be more or better than reality. This viewpoint influenced the perception of digital spaces as virtual spaces. Now nomadic technologies, smaller interfaces, and wireless sensors are embedding this virtual reality in public spaces, not because one is able to connect to the Internet while in movement, but because these interfaces re-define reality, expanding the emergence of possible and distant realities within the nearby space. It is important to stress that the disconnection between virtual and physical spaces contributed to the perception of an opposition between the concepts of virtual and physical. In newer hybrid realities, the virtual is conceived not as opposed to physical but as a potentiality already present in the physical. In this sense, virtual represents a broader aspect of reality.

Exploring this sense of virtuality, Gilles Deleuze (Deleuze 212) suggests that differentiation is creation. Differentiation is synonymous with actualization, since in the movement from virtual to actual an idea or a concept can potentially be differentiated into several actual instances. The potential to be actualized and differentiated into diverse realities is what makes the virtual an important part of the real rather than opposed to it. *"The virtual is fully real in so far as it is virtual,"* says Deleuze (Deleuze 208). The virtual is always ready to emerge, wishing to have actual existence. Deleuze focuses on the process of actualization as acts of differentiation, genesis, or creation. In this sense, artworks can be perceived as incarnations of potential ideas and desires, manifestations of potential structures. According to him, the act of creation in art occurs not between two actuals but between the virtual and its actualization.

The movement from virtual to actual (actualization) can be used to think about media artworks as virtual pieces. Every artwork can be envisioned as a virtual piece, which is differently actualized by each viewer. However, interactive pieces *per se* are fully potential entities that can only be completed when the interaction with the user occurs. Each user, in turn, actualizes the artwork in a different way, revealing some (but not all) aspects of its potentiality. This interaction can often be accomplished by the physical presence of the visitor and sometimes by her remote presence. Brazilian Professor Andre Parente (Parente 14) defines the virtual as a desire to constitute the real as new. In this context Deleuze, as well as Félix Guattari, Pierre Lévy, and Jean-Louis Weissberg, consider the virtual as a function of the creative imagination as well as a product of different articulations among art, technology and science. Therefore the virtual is capable of creating new conditions to model the subject and the world.

The emergence of hybrid spaces changed the relationship between reality and its simulation, and therefore the articulation between real and imagined spaces. As we have mentioned, the Internet has been regarded as a virtual simulated space, in which users would project imaginary and utopian spaces, mainly when inhabiting multiuser environments and creating new types of communities. Hybrid spaces bring these communication places again into urban spaces, but do they also bring imaginary spaces formerly projected onto digital spaces? How are imaginary spaces re-defined today with the emergence of a hybrid reality that merges actual and potential states? A brief glance at the world around us suggests that nanotechnology has an active role in constructing new imaginary spaces.

nano, for example, combines real and imaginary spaces, representing the world of nanotechnology through art and science fiction. *nano* not only blurs the borders between what is real and what can be imagined; the show also re-defines imaginary spaces by changing their traditional location. Imaginary spaces have historically being created outside the borders of physical and known spaces. Science fiction works about nanotechnology bring imaginary spaces as (unknown) folds within the known space.

4.2. Re-defining places of imagination: physical, virtual, nano

The projection of imaginary (inner) spaces onto external reality by means of art and narratives is as old as the human culture. Throughout history these projections of possible (or virtual) realities have been redefined many times. Although there are numerous sources of imagined realities, generally they have been connected to one idea: the existence of unknown and distant worlds. Whereas in former times imaginary spaces were located *outside* known and familiar space, they now move *deeper into* known space, which contains within itself the invisible nanoworld.

In order to better understand why imaginary spaces are today seen as inhabiting known spaces, it is helpful briefly to map the successive displacements of the imaginary from the physical to the digital and then to the nanoworld. During the Middle Ages, much of the popular imaginary was based on travelers' tales. Travelers went to distant and unknown places, which generally had no precise geographical position, and then returned to narrate their experiences. The construction of the imaginary has always had a close connection to the definition of borders, that is, to what is inside and outside the known space.

Travelers' tales were considered valuable not necessarily because they were accepted as literally true but because they stimulated the imagination. Italo Calvino starts the narrative of *Invisible Cities* as follows:

> "Kublai Khan does not necessarily believe everything Marco Polo says when he describes
> the cities visited on his expeditions, but the emperor of the Tartars does continue listening
> to the young Venetian with greater attention and curiosity than he shows any other messenger or explorer of his." (Calvino 5)

Because Polo represented the outsider, the one who came from a distant and unknown land, the veracity of the narrative was irrelevant as long as it nourished the Great Khan's imagination. It achieved reality through affirmation. The veracity of imaginary places constructed through the mediation of travelers' tales can thus be considered as mediated spaces. Imaginary spaces are frequently created when one is not able physically to access them. Mediating interfaces, standing at the borders of real and imaginary spaces, are critical to building these imagined (and often imaginary) realms.

As the globe has been increasingly mapped in more and more detail, imaginary spaces have been successively relocated. Science fiction illustrates how the space of the imaginary was moved from Earth to outer space. The work of Edgar Rice Burroughs, creator of Tarzan, exemplifies this trend. The Tarzan novels were conceivable at the beginning of the twentieth century because Africa was still not completely explored by Europeans. Because it was unknown, it served as a space for the projection of the imagination. But once Africa had been completely mapped, Burroughs stopped the Tarzan series and started writing about the planet Venus as an imaginary space. By the twenty-first century almost all planets of the solar system have been explored, and we are fairly sure there is no life close to us (with the possible exception of Europa). So where do imaginary spaces open up in the twenty-first century?

During the last twenty years cyberspace has frequently been regarded as the place where the projection of imagination could occur. The digital space represented a non-place, or a space located outside the borders of physical space. The idea of immateriality was critical to defining cyberspace as an imaginary place. Because it has been considered immaterial, it would be free from the constraints of the physical world. Therefore, it would be possible to create new places and new identities. It would be feasible to lose one's material body and still travel around the world as a body of information. In contrast to the Deleuzian idea of the virtual as potential, in the rhetoric of cyberspace the virtual has often been regarded as simulation: immaterial, non-physical, and non-real rather than as emergent potentiality. Several science fiction works in the past two decades have helped to project imaginary spaces onto cyberspace. William Gibson's trilogy, including *Neuromancer* (1985), *Count Zero* (1986), and *Mona Lisa Overdrive* (1988) made cyberspace a household name. Neal Stephenson's *Snow Crash* (1992) also helped to popularize virtual spaces.

Due to the increasing number of websites and cyberspace's commercialization, the desire of freedom does not quite fit what the Internet has become today. Nowadays, we can perceive a migration from cyberspace to nanotechnology, as authors become interested in representing the really small. In such works as Neal Stephenson's *The Diamond Age* (1995) and Michael Crichton's *Prey* (2002), nanotechnology becomes a territory waiting to be explored, albeit within known and inhabited spaces. Not coincidentally, both these works associate nanotechnology with the exploration of mysterious spaces at the margins of cities or densely populated areas; for *Prey* it is a nearly uninhabited desert, and for *The Diamond Age* the underwater realm of the Drummers.

The fact that these marginal spaces are associated with nanotechnology indicates that there is a possibility of creating the unknown even within the known. It is no longer necessary to travel to strange lands or to transport into cyberspace to find the unknown; it is folded within the known objects and spaces we inhabit in our everyday lives. Nano particles, generally not well understood by the general populace, are invisible even to scientists. One of the mechanisms to "visualize" atoms is the scanning tunneling microscope (SMT). In the STM's operation, a tunneling current flows when a sharp tip approaches a conducting surface of atoms at a distance of approximately one nanometer. The tip movement over the atoms and molecules is recorded and the data can be used to construct an image of the surface topography. At constant current flow, each individual atom on a surface can be resolved and displayed. Metaphorically, it is like a blind person who only knows the world by feeling its surface and can therefore imagine how surfaces would appear.

Interestingly, attempts to actually *see* atoms through the Transmission Electron Microscope (TEM) sometimes fail because some molecules cannot support the high energy levels that come from the microscope's electrons, "burning" the molecules. The TEM works much like a light microscope. However, the difference is the source of illumination: whereas the light microscope uses a beam of light, the electron microscope uses a beam of electrons to illuminate the sample. The wavelength of electrons, which is much smaller than the wavelength of visible light, sometimes destroys the tiny particles it attempts to illuminate.

Because nanotechnology is quite new and not generally understood, it has become an important source for the projection of imaginary spaces. Humans have always had a difficult time trying to understand what appears not to follow the "normal" course of nature (beasts and weird races), as well as what is not visible. Nanoscience encompasses both, since in the nano world particles are not visible to the human eye and behave differently than large-scaled matter. The *nano* exhibit speaks to our desire to know imaginary spaces by representing cellular patterns and mapping the invisible using sounds and graphics.

4.3. *nano:* representing the hybrid space

The creative team working at LACMALab on the *nano* exhibit created hybrid spaces that would reveal the interplay between actual and virtual realities. At sub-atomic levels particles pop in and out of existence, surface boundaries are dynamically unstable, and the observer affects what is observed. *nano* takes advantage of art mediated by technology to represent the universe of potentialities discovered and explored by nanosciences. Connecting nanotechnology to imagination does not imply that the science itself is imaginary; rather, it is related to how people project their imaginaries onto a potential and unexplored part of the real.

As previously discussed, the emergence of media arts is responsible for reconfiguring museums as public spaces, creating more interactive and participatory spaces. The museum space becomes a hybrid environment encompassing virtual and actual realities. *nano* creates a hybrid reality by allowing remote visitors to use physical avatars *(robotic spheres)*, merging physical architecture with projections that represent "invisible" realms (floor projection grid), and treating human beings as quantum particles that interfere in non-contiguous spaces *(Quantum Tunnel)*. Most of all, it creates a hybrid space because it merges potential and actual spaces.

The hybrid space starts to configure as soon the visitor enters the exhibition. The placement of the visitor's picture inside the virtual projection on the wall, mixed with other faces that have been there before but are absent now, is the first sign of hybridization. The mixing of virtuality and actuality is strengthened by text passages projected on the wall. "You are the sum total of your data,"(DeLillo, *White Noise*, 141) a text passage close to the camera swarm, asserts that human organisms are "digital/genetic" data.

Sense Spaces invite the visitor into the invisible realm of nanotechnology by creating an immersive sensation that is mainly evoked by audible and tactile experiences. According to Andrew Pelling, a nanoscientist graduate student co-responsible for the sound, *nano* aims to make sound itself a tactile experience. Part of Andrew's research focuses on converting the ASCII data obtained from oscillatory (beating) movement of cells sensed with the Atomic Force Microscope (AFM) into sounds. Nanobots (nano-sized robots) are often envisioned as armies of microscopic machines. Similarly, inside a cell millions of proteins are "swarming" to make things happen. Therefore, in *Sense Spaces* small sound bits and words come together and swarm, contrasting with large moments of silence and "background" noise.

The *Inner Cell* works with sound in a similar way. Sub-woofers are used to make the visitor feel the bass associated with the floor, so that the floor surface pulses and responds to movement by people and *robotic spheres* in its space. The floor projection, an analogy to nano-space, is composed of a grid representing a hexagonal pattern of graphite carbon atoms. In the nano-world every particle influences each other, so the space is deformed by the presence of other particles, in this case visitors and *robotic spheres*. The floor has a wave-like behavior, similar to what happens in the nanoworld. When the STM scans atoms, the data is used to produce visualizations. The wave-like behaviors revealed by the visualizations are caused by electrons, which can behave like waves as well as particles. Therefore they are dispersed in space, and this nano-dispersion is represented by wave-like appearance. Every nano-particle that comes close enough to an atomic surface affects it and is affected by it. Causing perturbations in the floor projection, visitors are like atoms experiencing quantum mechanical interactions.

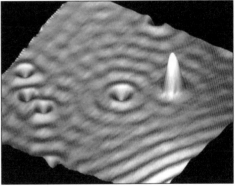

Sketch of *Inner Cell,* by Ashok Sukumaran

PicoLab, UCLA, Lisa Wesoloski, Shane Dultz and Jim Gimzewski

The graphite floor and the scanning of graphite atoms with a STM.

The *Inner Cell* also mixes the actual architectural space with potential representations of cellular movements. The interaction with the buckyball projection includes a double sense of the virtual. Molecular structures resembling the buckminsterfullerene C-60 carbon molecule are born on the virtual space of the wall, and visitors can manipulate the molecules with their shadows. The piece acquires meaning through interaction with visitors. The molecules can be moved and squeezed in different ways, creating interactions in which visitors and atoms mutually influence each other.

The C-60 molecule represented by the piece was discovered in 1985 in an experiment to unravel the carbon chemistry in red giant stars. Until then, carbon was known in the form of diamond and graphite. Whereas diamond has a modified tetrahedral structure, graphite forms a flat grid composed of hexagons. In contrast, the buckminsterfullerene carbon-atoms are connected with a slight angle in between, creating a closed cage structure that resembles a soccer ball formed by 12 pentagons and 20 hexagons. The deformation of the buckyballs in the piece is similar to what happens when an STM pushes the C-60 molecule using electrons that come from the microscopic tip. Electrons flow gently through the tip, and if this flow is monitored, it is possible to see the deformations created by the tip literally "poking" the buckyball. In the projection, the visitor's shadow works as a tip that can deform and move the buckyball. Only slow and smooth movements can move the molecules, indicating that fast and rough movements are not (yet) effective at the atomic level while manipulating atoms. In playing with scale, the piece aims to make people aware of how atoms and molecules behave at this invisible level of reality.

Finally, the *Inner Cell* is inhabited by *robotic spheres*. Representing giant atoms, these spheres embody virtual participants, inverting the traditional meaning of a digital avatar. The Hindu word avatar, widely used among the digital community, designates a being who is the embodiment of the god Vishnu. In digital parlance, when one chooses a virtual character to represent oneself, this "creature" becomes one's avatar, so that an avatar is a digital representation of a physical body. In the case of *robotic spheres* this relation is inverted, since the *spheres* have actual existence that respond to actions of individuals who are not physically present in the *Inner Cell*. The initial idea behind the *robotic spheres* was to create a remote website through which they could be controlled. Later, the control interfaces moved to the *Atomic Manipulation Space*, allowing participants in that exhibit module to interact with the adjacent *Inner Cell* module. In both cases the idea is to combine remote and contiguous spaces into a hybrid environment and also to have visitors interacting with physical avatars. Manipulating the *spheres* also affects the visitors in the *Inner Cell*, because they feel compelled to move whenever a *robotic sphere* approaches. In this sense, visitors also have the power to manipulate and affect other visitors in the exhibition. The remote manipulation of the *spheres* has been inspired by the work of Donald Eigler and Erhard Schweizer who in 1989 spelled out "IBM" by individually arranging 35 xenon atoms onto a nickel surface with a STM.[13] The ability of remotely influencing and rearranging the spheres represents the idea behind (the imaginary of) nanotechnology in building matter from the bottom-up.

The same thought is represented in the *Nanomandala*. As mentioned before, the installation deals with the process of constructing our universe by connecting grains of sand as a metaphor for the basic structure of life to atoms as the building blocks of our universe. The *Nanomandala* also connects the LACMA East, a traditional museum where the original mandala constructed by the monks is exhibited, to the LACMALab, since it projects over the sand the original work in the adjacent museum. Following the purpose of traditional museums, visitors are not able to interact or touch the Mandala exhibited in the LACMA East. Conversely, the *Nanomandala* is designed to invite interaction, encouraging visitors to play with the sand, highlighting one of the characteristics of hybrid museums.

Hybrid spaces can also be understood as a dynamic enfolding of different contexts and scale levels into one another, via digital technology. Kaleidoscopic structures spread throughout the exhibition contribute to folding the space of the exhibition inside itself. When the visitor moves around *nano*, he or she is able to look at distorted views of the *Inner Cell* while being immersed in a soundscape of fictional narratives about nanoscience. The Kaleidoscopes play with the idea of looking *in* from the *outside*, creating an interplay between inner and outerspaces.

Lastly the *Quantum Tunnel*, like other parts of the exhibit, is based on the idea of potentiality and possibility. Quantum tunneling is described by quantum mechanics as the probability that an electron, when encountering an energy barrier, goes through it instead of bouncing back. There is a finite probability this could happen with large-scaled entities like human beings, although here the probability is so infinitesimal that it would not happen once during the lifetime of the universe. The *Quantum Tunnel* addresses human beings as if they were electrons tunneling from one space to another, capable of altering other people's particles. To illustrate this dynamic, imagine a traditional theater with a stage and characters. Now suppose that the stage is elastic and deforms itself depending on the characters' movement. It is as if the stage is also a character, interacting with other characters.

The distinction between the actual and virtual blurs in the nanoworld, since electrons have no precise physical location, can behave like particles or waves, and can apparently jump from one point to another without moving through the intervening space. Nano-particles are simultaneously potential and actual entities, revealing the merging of these states into a hybrid reality. How can anything be considered as strictly *either* potential *or* actual in the nano-world?

5. Conclusion: blurring borders between real and imagination — potential futures

The blurring of potentiality and actuality in the nanoworld, and the lack of general knowledge about nanotechnology, create a fertile imaginary around the new discipline. A common nightmare speculates that, with the aid of nanotechnology, researchers will build nanostructures capable of replicating themselves like nano-robots. UCLA Professor James Gimzewski relates that when he worked at IBM "a newspaper called the *Bild* printed a front page story saying 'IBM creates nanobots that can cure cancer' with a picture of them swimming inside the human body and describing it as having a cancer-killing unit that used lasers to 'blast away' the cancer cells."[14] Immediately, there were people from all over the world calling IBM and asking how to get these nano-bots.

The nano-bot story was not true, but there are many future developments for nanotechnology that might potentially have a great impact on our future. Several possible inventions aim to develop bio-chemical sensors responsive to the environment. For example, windows can cool the ambient air if it is too warm outside, and clothes can warm the wearer if it is cold or cool one down if it is warm. Another application envisioned by Professor Gimzewski is the use of biodetectors in restaurants "and in any public place, which can be used by uneducated people and that will detect the presence of viruses and different types of hazardous material."[15] Nanotechnology can also be used to engineer intelligent drug release systems and to manipulate cell structures inside the human body. This idea has been explored by science fiction for some time. For example, the 1985 movie *Innerspace* narrates the story of a group of people who are miniaturized and injected inside the body of a hypochondriac. Most of the dreams and nightmares related to nanotechnology are connected to the creation of nano-structures and molecular machines that could go inside the human body in order to fix or destroy it.[16] Other favorite science fiction themes include the creation of pervasive surveillance and invasive devices that could be injected inside the body. What if somebody could control a person's dreams without any visible interface?

Mark Weiser used to say in the last decade that ubiquitous interfaces could be considered "natural" because their existence would not be perceived.[17] The smaller the interface, the more "natural" it is. With nanotechnology some interfaces are no longer perceivable by the human eye, and so will "naturally" become an accepted part of the environment. In this context, media arts play important roles in making people aware of new technological innovations. Art has always been concerned with representing imaginary worlds, and it has also often pushed the limits of technology to change the physical world around it. Initiatives like the *nano* exhibit, and more broadly the shows commissioned by LACMALab and other such experimental venues, foment this type of discussion, rethinking the role of art and technology and re-defining the borders between real and imaginary spaces. The dialectic now takes place between the actual and the virtual, both of which are participating in constructing our reality. In working with the in-between space that connects art and technology, *nano* and LACMALab create a hybrid space that, without being didactic, enlarges the scope of the real to include what potentially can be as well as what actually is.

6. The visitor glances back

From the *Studio Area*, I look at the wall and read in very big letters: "Nature is Imagination."[18] Watching visitors at the Boone's Children Gallery, I think that the museum is no longer the same, and neither are our imaginative constructions of the spaces in which we live.

○ ◇ ○

nano

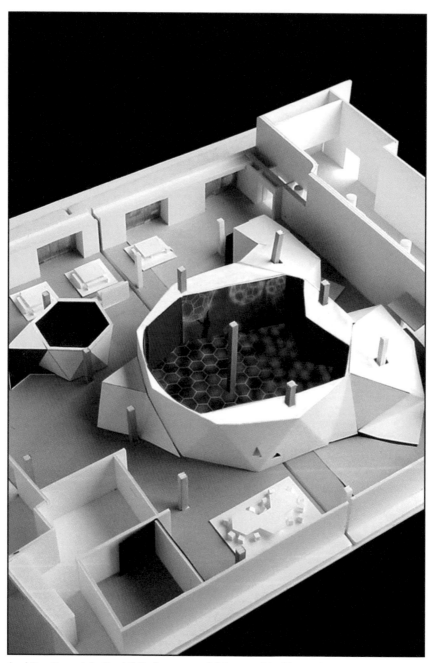

Architect's model of exhibit. Courtesy of Johnston Marklee & Associates.

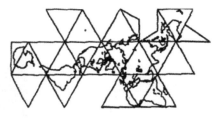

Buckminster Fuller's Dymaxiam Map inspired the exhibit's architecture.
Courtesy of Johnston Marklee & Associates.

Surveillance cameras capture visitors' images and
encapsulate them in self-assembling hexagons.
Courtesy of Michael Chu.

Both images: In the *Inner Cell*, the zero@wavefunction interactive projection shows carbon-60 molecules, familiarly known as buckyballs. The images can be manipulated using a visitor's shadow. Courtesy of Jiacong Yan.

Entrance/exit to Inner Cell with embedded sounds of text quotations.
Courtesy of Nano Creative Group.

Robot spheres in Inner Cell.
Courtesy of Nano Creative Group.

Nano Creative Group members take a break on the hexagonal interactive floor projection of the Inner Cell. Courtesy of Jiacong Yan.

The Atomic Manipulation table with track balls that control robot
spheres in the Inner Cell. Courtesy of Nano Creative Group.

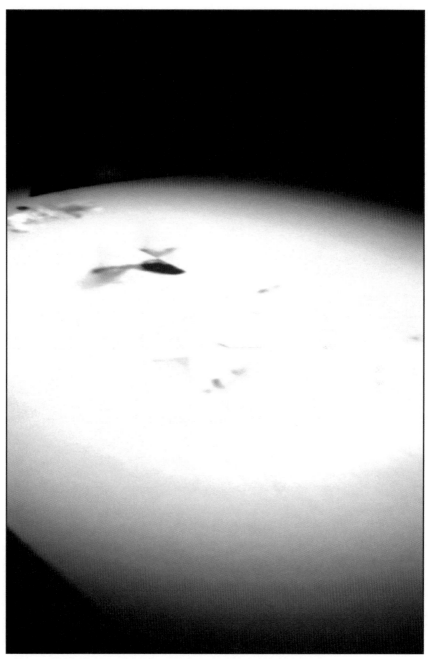

Image of visitors and robot spheres on the Atomic Manipulation table.
Courtesy of Nano Creative Group.

Exterior of the Inner Cell with kaleidoscopes looking into the interior.
Courtesy of Nano Creative Group.

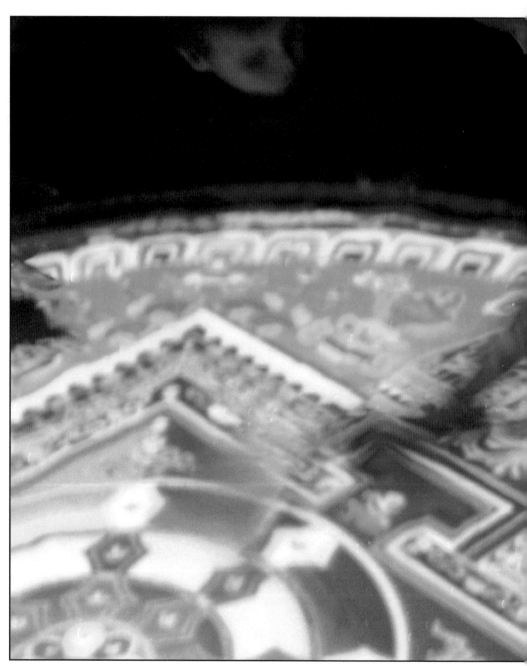

Nanomandala projected on sand-filled table.
Courtesy of Jiacong Yan.

Center image of nanomandala created by Buddhist monks.
Courtesy of Jiacong Yan.

Individual grains of sand in nanomandala.
Courtesy of Jiacong Yan.

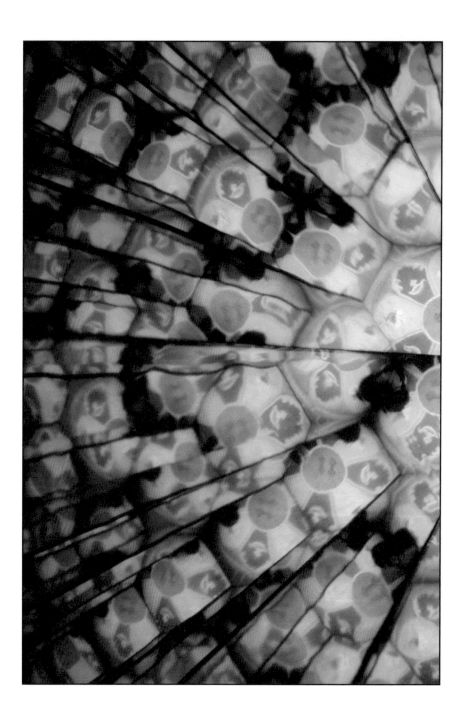

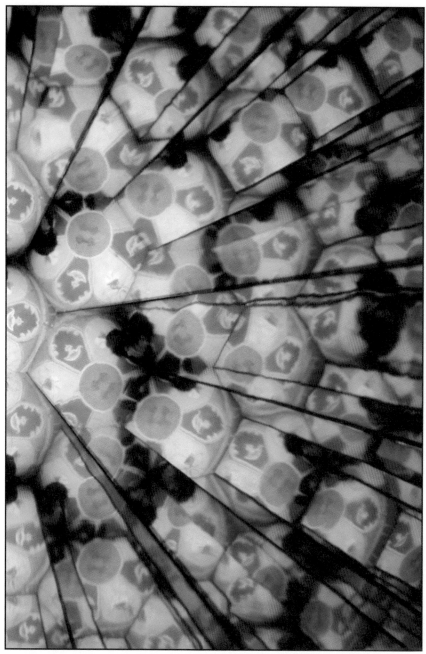

Kaleidoscope images with live feed image from nanomandala.
Courtesy of Jiacong Yan.

Video camera stalk in Quantum Tunnel.
Courtesy of Ashok Sukumaran.

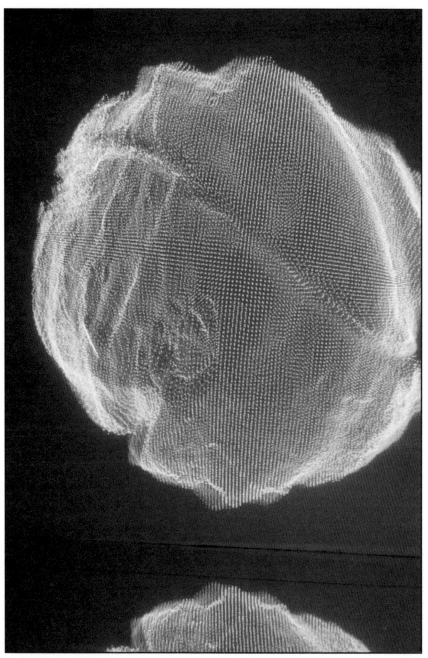

Projected wall image from video camera in Quantum Tunnel.
Courtesy of Ashok Sukumaran.

Fluid Bodies projected text dissipating as visitor walks by.
Courtesy of Nano Creative Group.

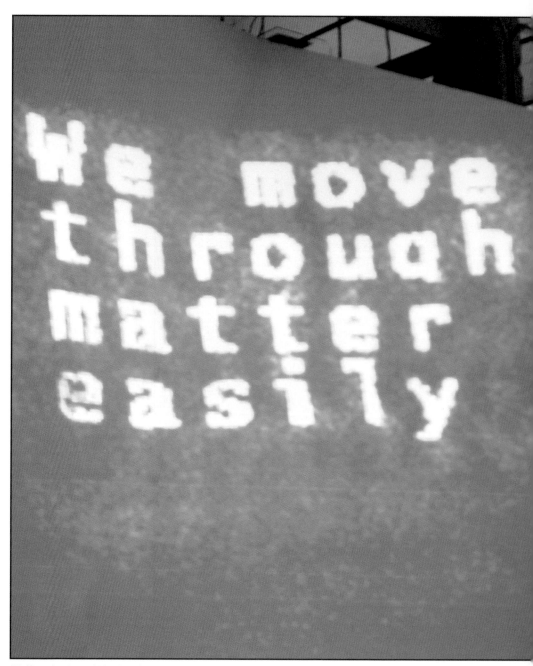

Wall projection of Fluid Bodies with text intact.
Courtesy of Nano Creative Group.

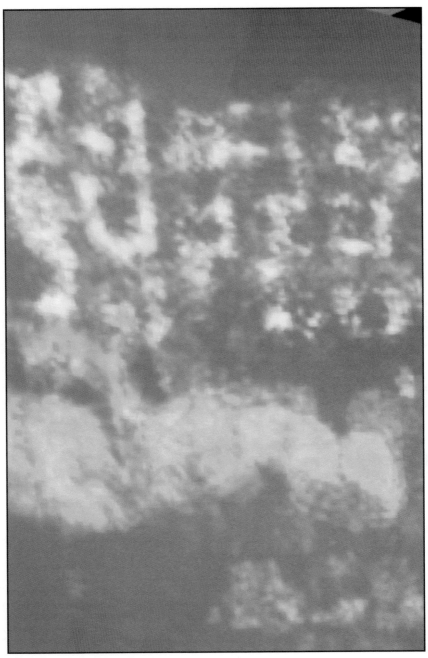

Fluid Bodies text dispersed through visitor interaction.
Courtesy of Nano Creative Group.

Crystal Method by Steven Schkolne showing shape drawn in space that can
be grabbed and rotated using sensing tool. Courtesy of Nano Creative Group.

Resource area with molecular models, books, and other materials.
Courtesy of Nano Creative Group.

Working Boundaries on the *nano* Exhibition
CAROL ANN WALD

Interdisciplinarity and Collaboration

When Katherine Hayles asked me to contribute an essay to this collection, I had recently finished six months of participant-observation research at a university robotics laboratory. One of the things I most admired and envied about the lab was the highly productive and supportive collaborative community that the lab director fostered among graduate students, post-doctoral students, and faculty. Observing and writing about the process of interdisciplinary collaboration on the *nano* exhibition project seemed a natural choice for an essay topic.

My aim, then, is to unpack the term "interdisciplinary collaboration" as it has played out over the eight months I have been observing the *nano* project.[1] Both halves of this term break up into multiple, sometimes contradictory meanings, in theory and in practice. The term "interdisciplinary" has been theorized and studied in numerous realms, from science to business, by social scientists as well as by scholars in the humanities. This essay will draw in particular on the concept of boundary work articulated by Julie Thompson Klein in her landmark examination of research on academic interdisciplinarity *Crossing Boundaries: Knowledge, Disciplinarities, and Interdisciplinarities.*[2]

This essay tells a story about some of the ways in which project principals fruitfully crossed and blurred boundaries between disciplines, embarking on mutually transformative dialogues, as well as ways in which divergent disciplinary enculturation sometimes created dissonances and sparked defense of boundaries. At several critical points during the project, objects and spaces functioned as border zones where different disciplinarities and differing understandings of "collaboration" interacted, often in charged emotional circumstances. Collaborative style often proved tightly coupled with disciplinary ethos.

I focused my analysis primarily on relations among the principals, but there is another important story to be told. This other story concerns the very different nature of the collaboration among students, who performed most of the actual labor. Art and science students worked together day and night for many months to create the exhibition. Along the way, they forged mutually satisfying, even transformative, professional relationships and friendships, in which disciplinary differences did not matter as much as shared interests, and goals, and work styles. They went about their tasks in a practical manner, regardless of conflicts among the principals. A clear difference between these stories is the degree to which power relations became entangled with the dynamics of disciplinary cultures and collaboration.

Positions, Players, and Places

Positions

Contemporary critiques of objectivity in science studies and anthropology—the recognition that the situated and partial nature of the observer's perspective both constructs and constrains knowledge—demand that I examine my own subject position in relation to this research. My allegiances and training make my position a tangled one. Katherine Hayles, a principal collaborator on the *nano* exhibition and the editor of this volume, is directing my dissertation in the UCLA Department of English. As instructor and mentor, she has played a central role in shaping my intellectual life. During my time observing the *nano* project, I spent more time talking with her than with anyone else on the team. The many strands of my relationship with her represent the most direct influence on my view of events. My training in literature, science studies, and cultural studies entwines about this relationship, ineluctably framing my point of view in my own disciplinary terms. I, too, am engaged in boundary work on multiple frontiers.

Money, as well as direct participation in the shaping of the exhibition, further complicated my position. During the months that I observed the project, I was employed both by Hayles and by another principal collaborator on the *nano* exhibition, Victoria Vesna. At the same time, I participated in some aspects of the creation of the *nano* exhibition itself, chiefly by contributing suggestions for text passages to be used in the project. I wrote several drafts of a fundraising brochure. At meetings I often put down my notebook and pen to take part in lively discussions about various aspects of the exhibition.

As an observer, my research consisted primarily of attending meetings and interviewing project members. I sat in on a number of team gatherings and retreats at UCLA, as well as some of the meetings at LACMA between the eight project principals. I interviewed seven of these, along with a number of students from the three participating UCLA departments.[3] Along the way, I continually engaged in informal interactions with many team members. Practical limitations on my ability to spend time observing team activities, such as the almost full-time daily labor of art and science students on design and construction of the exhibition, inevitably limited my view of the project's full collaborative richness.

Beyond these many constraints on my perspective, as I wrote this essay I realized that Hayles' role as an informant conflicted irreconcilably with her role as my editor. As a participant, she had her own perspective, as did all the principals. On the other hand, as an editor, she expressed a commitment to having all viewpoints represented in my essay. The charged problem of Hayles' double role permeates my writing like an alternating electrical current.

Players

The exhibition team has grown substantially since the first brainstorming meetings in January of 2003. The principal UCLA collaborators were the Chair of the Design | Media Arts program, Victoria Vesna; Professor of Chemistry and nanoscience pioneer James Gimzewski; and Professor of English and Design | Media Arts Katherine Hayles. Robert Sain, director of LACMALab, the Los Angeles County Museum of Art's (LACMA) ongoing experiment in participatory museum experiences, and Carol Eliel, LACMA's curator of Modern and Contemporary Art, were the principal players on the museum side. Sharon Johnston and Mark Lee of Johnston Marklee & Associates joined the project principals as architects, primarily represented at meetings by Johnston and her associate, Anne Rosenberg. While Williamson, a freelance science writer, was asked to become a principal collaborator at Sain's invitation, according to Gimzewski, she actually attended only a few meetings. Vesna, Gimzewski and Hayles each brought students into the project at the beginning. Over time, the team has grown to include numerous others, including outsourcing firms, graphic designers, a web designer, computer technicians, lighting technicians, carpenters, painters, and many, many more.

Places

As Klein points out, boundary is a spatial metaphor.[4] Places are important, for control of physical territory helps to determine power relations: Who has the ability to provide spaces for work and meetings? On whose turf do activities occur? As in team sports, playing at home can have advantages. In January 2003, Vesna, Gimzewski and Hayles invited a group of students to a three-day "synaptic blow-out" brainstorming session.[5] Each day, the group met at a different location: Hayles' home in Topanga Canyon; the Experimental Digital Arts room (also known as the EDA: a large, reconfigurable meeting, exhibition and lecture space in the Design | Media Arts department); and finally in Gimzewski's lab.

Subsequently, as the exhibition began to take shape, regular weekly meetings took place in the EDA. This room became one of several staging areas for the exhibition, preparatory for the move to the Boone Gallery at LACMA in the late fall of 2003. The group enjoyed monthly, daylong retreats at May's Landing, a Malibu house owned by UCLA, where intensive and productive meetings on the patio were punctuated by potluck lunches and beach walks. Meetings with LACMA staff and the architects usually took place either in a LACMA conference room, or in the Boone Gallery itself; the LACMA contingent occasionally joined us at May's Landing. Work on the design and construction of key elements of exhibition modules was distributed throughout several locations: the EDA; a computer laboratory called the Creative Technologies Lab; Vesna's studio space; and a laboratory under the direction of Gimzewski called the PicoLab. The architects' studio in West Los Angeles was also a key locus of collaboration. Here, apart from the others, Johnston and Lee engaged in intense discussions about how to shape the physical infrastructure of the exhibition, and created their drawings, diagrams, and models.

Boundary Work and Notions of Interdisciplinarity

"Boundary crossing," says Klein, "has become the defining characteristic of the age."[6] Yet, paradoxically, while knowledge is "increasingly interdisciplinary," "the long-term structural trend of academic institutions has been in the direction of greater specialization, professionalization, departmentalization, and fragmentation."[7] It is important to note that much of the real work in the world is performed on a project-oriented basis, in which disciplines do not matter nearly as much as they do in academia. Even in some academic institutions, innovations have been introduced that reorganize teaching and research around project-based models. The disciplinary and professional positions of the three UCLA scholars who are principals of the *nano* exhibition encompass all of these trends.

Internationally known as a media artist, Victoria Vesna came to UCLA several years ago to head up a new art program that emphasizes electronic and digital media, a program different from the more traditional studio arts specializations at the university.

Soon after she took up her position at UCLA, Vesna invited Gimzewski to participate in a panel entitled "Nanotechnology and Culture" at a conference entitled "Networks to NanoSystems," which she co-organized in 2001. He was also an internationally known scholar who had recently come to UCLA to head up a nanoscience laboratory in the chemistry department, just as the university was breaking ground on a new, well-funded interdisciplinary nanotechnology research unit, the California Nanosystems Institute. He had originally worked in chemistry, then became a pioneer in the new field now called nanoscience. Gimzewski had always been interested in art and creativity; Vesna had always been interested in science, but had chosen to focus on art. They shared a drive to push their work beyond the traditional boundaries of their respective fields.

As they became friends, their conversations began to generate ideas for collaborative art projects on nanoscience themes. Their working relationship was already firmly established before they began the *nano* exhibition project, and their first collaboration, *Zero @ Wave Function: Nano Dreams and Nightmares,*[8] formed the kernel around which NANO grew—the fruit of a long-running conversation between Vesna and LACMALab's Robert Sain about the role of digital technology in the future of the museum. Reliance on electronic and digital technology increasingly links scientists and artists. In their jointly edited anthology *Art @ Science,* longtime collaborators Christa Sommerer and Laurent Mignonneau argue that digital technology, in particular graphical computer science, has the potential "to make connections among a wide variety of arts and sciences." They speak of a new spirit of holism, arguing that "art and science should no longer be considered separate and contrary disciplines," and that "interactive technologies" are catalyzing this synthesis.[9] Vesna and Gimzewski's work together reflects this theory.

Implicit in this holistic position is the belief that computing and new visualization technologies are inevitably causing widespread change in the arts and sciences.

Zero @ Wave Function illustrates how digital visualization technologies are constructing new common ground between scientists, computer scientists, and new media artists. On *nano,* a common interest in visualization technologies helped to create strong bonds between Gimzewski's students and Vesna's students. A few of the literature students whose interests encompassed such topics as visual poetry and electronic literature also connected with the art and science students on the common ground of new visualization media.

Science studies is another relatively new, interdisciplinary field, encompassing work in literature, anthropology, sociology, philosophy, history, and women's studies. Katherine Hayles originally trained as a chemist, but switched to literature and has become a well-known scholar in the area of literature and science. In the last few years, Hayles has developed an interest in electronic literature and new media arts. After getting to know Vesna at a D | MA lecture series, Hayles team-taught a course in her department and eventually took on a **joint** appointment in Vesna's program.[10]

Project principals staked out different positions on the possibility of crossing disciplinary boundaries. These differences came into focus around the notion of "transdisciplinarity." "Transdisciplinary" generally refers to "a paradigm or vision that transcends narrow disciplinary worldviews through overarching synthesis." Transdisciplinarity is an idea that looks back to "older notions of unity and simplicity, as well as new searches for coherence and connection in the modern world."[11]

Hayles expressed skepticism about this construction of transdisciplinarity:

> I don't believe in the transdisciplinary idea. I think that our training in disciplines has a profound, profound effect on how we think about intellectual problems, how we approach them, how we solve them or don't solve them, and I don't think it's possible and maybe it's not even desirable to overcome that.

She described two other views of transdisciplinarity that fall short of overarching, synthetic theory:

> One vision of transdisciplinary work is that we somehow transcend our disciplinary basis and we're able to look at a project, a problem from a broader perspective that takes into account several different disciplinary orientations and from this elevated—I would say almost superhuman—perspective be able to address the problem. Now there may be a true polymath like Buckminster Fuller or Stanislaw Lem who is so brilliant and so creative and original that they can in fact do this, but I think it's far beyond the scope of most people because it involves a deep and real understanding of really different knowledge bases. …The other definition of transdisciplinary which I think is very difficult but still more possible than the other, is to be able not only to do work that addresses two or more fields, but actually to make a contribution to both fields. … [Y]ou're changing the understanding in both disciplines about a common problem. And that is very difficult to do, because it implies that you have a sufficient knowledge base to really address substantively practices in two significantly different disciplines.

Hayles points to physicist Erwin Schrödinger's book *What is Life?* as an important instance of this mode of transdisciplinarity: Schrödinger's recasting of biological problems from a physicist's perspective profoundly influenced how biologists look at their field, and motivated physicists to turn to biological problems.

Vesna, unlike Hayles, employs the term "transdisciplinary" to describe the kind of work to which she aspires:

> 'Inter-' basically [denotes] 'international' …'in between': two separate things. As long as we see things, people, disciplines as separate, we're thinking [in terms of] 'inter-.' When we start thinking [in terms of] 'trans-,' we're actually thinking beyond disciplines. I have …taken all the 'inters' out of my lectures and speeches and moved to 'trans-'.

For Vesna, transdisciplinarity means "not amplifying an existing mode but cutting it off at the root and floating free of disciplinary boundaries." Gimzewski also often speaks about the need to move beyond disciplinarity. He and Vesna imagine a future educational system in which students are trained to think and work simultaneously as artists and as scientists.

One of Gimzewski's students eloquently testified to the promise of this approach. He spoke of the ways that current academic institutional structures do not grant much room for scholars, like Gimzewski and Vesna, who try to work in new ways not traditionally sanctioned by their disciplines. For this student and others, Gimzewski and Vesna have pointed toward a liberating new path for artists and scientists who want to cross and blur disciplinary boundaries to work together.

The views of Vesna and Gimzewski on transdisciplinarity are increasingly entering mainstream scientific research discourse. In 1998, Professor Heidi Diggelmann, president of the Swiss National Science Foundation, argued the usefulness of transdisciplinary approaches in solving real-world problems such as climate change. At the same time, she underscored the need for foundational disciplinary knowledge:

> There can be no doubt that research perspectives of this kind open up new channels for generating and transferring knowledge. But this does not make the individual subjects and disciplines obsolete. On the contrary, a solid grounding in the disciplines and scientific excellence are vital preconditions for high-quality trans- and interdisciplinary work. The watchword is not "either/or" but "both."

Capturing Collaboration

Vesna, Gimzewski, and Hayles possess clear notions about the nature of interdisciplinarity. How do individuals from different disciplines envision collaboration? When I asked project principals to describe their collaborative ideals, they spoke first about the necessity of trust and respect for each other's expertise and areas of competence. Respect for differences in competencies led to divisions of labor, but did not preclude misunderstandings about these competencies, or struggles over questions of creative control. Through dialogues that ranged from collegial to tense in tone, the principals attempted to reach some degree of understanding of each other's concepts about collaboration that would allow work to progress. They accounted for disagreements about the nature of collaboration by employing a range of explanations: the practical necessity of leadership; degrees of authority based on contributions of resources; the incompatibility of work cultures; the importance of a holistic approach to the project; and the primacy of disciplinary expertise.

The creative process that Robert Sain desired for nano departed significantly from that of his earlier exhibitions. In previous LACMALab shows, "Seeing" and "Made in California NOW," Sain and his assistant, Kelly Carney, working with a LACMA curator, had developed the themes and goals of the exhibitions, then brought in a number of individual, unrelated artists or groups to create installations linked by the LACMA team's concept. Nano was a radical departure from their previous way of working. For nano, Sain conceived the idea of assembling a large team to create an exhibition collectively. "What was being proposed here," said Sain, "was a true swirling together, whether you call it collective or collaborative, or even [a] 'nano' approach of tearing down ... boundaries."[15] He and Carol Eliel envisioned a transdisciplinary process that "gets rid of hierarchy" while simultaneously honoring differences in expertise. It was to be about building an environment "where everybody's a teacher and everybody's a learner." Eliel's account was similar: "Ideally the way we had envisioned it—and Bob and I felt we had clearly described [it to the other principals] from the beginning—was [that] the eight principals . . . all had equal voices in this collaborative process in terms of shaping the exhibition, that there was not a hierarchy in that group."[16] Despite Eliel and Sain's belief that they had made their vision of the process clear to the rest of the principals, Johnston, one of the architects, did not think Sain and Eliel communicated as well as they had asserted: "We did not feel the structure of all the team members was that clearly articulated at the outset, between horizontal and hierarchical roles and responsibilities."[17] This confusion about roles and hierarchy contributed to tensions among the principals.

In elaborating his ideal, Sain deployed the metaphor of companionate marriage, in which individuals are bound together by reciprocal ties of trust and respect that dissolve ego boundaries:

> It's not like we're all going to vote. . . . I think you don't necessarily have to have consensus all the time. I think it's not unlike any other kind of relationship. This is a leap, but in a marriage, is the goal consensus? Well, I guess it's easier if there's consensus... It's still more about building a relationship such that questions about consensus sort of evaporate.

When Vesna approached Gimzewski about expanding their previous work on *Zero @ Wave Function* into a much larger exhibition, they also dreamed of fashioning a new way of working—one that involved breaking down boundaries between disciplines and, by implication, between team members: "[I]mmediately we thought, we're not going to separate science from art from architecture from literature from gallery curator. We're going to create something truly unified," Gimzewski said. Pre-dating this aspiration, two other models of collaboration had already been established in Gimzewski's work life: the collaborative relationship between himself and Vesna, and the culture of the scientific research laboratory in which he had been immersed for many years.

After Gimzewski and Vesna met at "Networks to Nanosystems," they began a long-running dialogue that included teaching each other about their work, discovering common interests such as the theories of Buckminster Fuller; and bonding around a shared belief that both art and science are, fundamentally, about imagination, "wonderment," and creativity. They forged the kind of relationship of equals Sain described as ideal, in which each is both a teacher and a learner, said Gimzewski:

> I found what she was saying so stimulating . . . and she also was stimulated by me.... I learned so much from her. At the same time, she was learning a lot in terms of the holes in her education in the scientific aspects. So it was really a two-way flow.

Gimzewski felt that his dialogue with Vesna had been transformative: "What it did was eliminate obstacles in my imagination. It changed how I felt about my work." Exchanging ideas with Vesna led him to shift his focus from molecules, which he understood very well, to cells, which presented a much more challenging level of complexity:

> I wanted to take a dangerous jump. . . . Working with Victoria empowered me to take a leap and to not be so boxed in and influenced by my peers. . . . I feel I've changed through

knowing her. It's not only me that's changed, it's also my group that's changed. It rubs off on them, and they're willing to be much more adventurous experimentally.

Gimzewski found in Vesna an ideal collaborator: they shared the give-and-take of mutual respect between equals; their conversations resulted in a transformation of their conceptions of disciplinarity and of collaboration. Gimzewski's new intellectual boldness affected the way he interacted with students, although initially his model for collective work had been that of the science laboratory in which the lab director is a sort of independent dictator: "I'm used to being given a budget; I have complete control, and do whatever I want to do." Gimzewski felt he had to give up his old, more hierarchical and controlled way of working, and embrace Vesna's much more open-ended work style, in which the end product evolves over time: "At first I couldn't sit in a room and discuss the project, because I wanted a much more hierarchical system. . . . I realized that I believed 'here is the concept and this will be the result.'"[18] As he assimilated Vesna's looser work process, he fashioned a new metaphor for collaboration based on the notion of organicism:

> When architects don't just work in an architect's office, and chemists don't just work in a lab, but you put them together as people who have to communicate, they become themselves a living entity. They become like the difference between individual cells and conglomerations of cells, which make tissue. . . . I imagined a tidy collaboration which produced what I would like . . . and what we saw is organic growth. . . . For me at times it was quite frightening, because things were changing too much. And then I learned you have to let go and embrace change, learn from change, and use it. And that is beautiful, because that is more what people should do in nanotechnology. . . .

Gimzewski employed an organic metaphor for collaboration that encompassed many more than two people, and which blurred the boundaries between individuals. Hayles, in contrast, like Sain used a comparison based on a marriage between two people to describe her ideal collaboration. She developed the metaphor to include the idea of creative synergy:

> My conception of collaboration in an ideal sense is a little bit like an ideal marriage except you don't have to deal with dirty socks. That is, the joining of partners in a common endeavor that creates a kind of synergy where the sum total of their efforts is greater than just adding each one alone together. That something extra happens, and that something extra is the result of mutually stimulating and complementing one another.

Because writing is central to her discipline, she conceives of collaboration first in terms of two people writing together. The goal is the creation of a "third voice." By conceiving of collaboration as producing a "third voice," Hayles preserves boundaries between individuals working together. Collaborators acknowledge and respect each other's differing areas of competence, and make divisions of intellectual labor. Hayles used the example of her collaborative writing work with her husband, an anthropologist and computer scientist. As they discuss a problem of mutual interest, moments arrive when each defers to the other's authority on a point because they mutually respect one another's disciplinary knowledge and experience. While the "third voice" echoes the holistic, synthesizing aspect of transdisciplinary theory, difference remains constitutive for

Hayles. Vesna and Gimzewski conceive of disciplinarity as permeable zones, so that distinctions such as "artist" and "scientist" lose meaning.

At the outset, Hayles believed that somehow this model of collaboration as a marriage of two distinct equals could work when the number of partners increased:

> I'm not sure that I had really thought this through at the beginning. I did have a clear sense that Victoria, Jim and I each brought different expertise to the project. And the challenge would be to find ways to put this expertise together. But I did sense the clear potential for a kind of complementary relationship because of the fact that we each knew different things, and therefore I could see how all these things could go into creating a successful exhibit.

It is worth lingering on Hayles' and Sain's marriage metaphors to try to understand some of the dynamics of collaboration among the principals. The conventional ideals of Western marriage delineate a relationship of trust, respect, and companionship between two people who are free and equal individuals. Johnston and Lee, partners in life as well as in profession, enjoyed such a relationship. Although living with different partners, Vesna and Gimzewski shared this kind of connection on an intellectual level. They were close friends who shared many beliefs, attitudes and interests. They had grown to trust each other implicitly. They had gained "communicative competence" across disciplinary divides by working together on a previous art/science project. Vesna always found it "easier for me personally to find a language and connect processes with the science group."

Vesna and Gimzewski also emphasized their shared approaches toward innovation and risk, and a mutual belief that some sort of commonality exists between creativity in science and creativity in art: "For the first time I felt genuine interest [from Gimzewski] in pursuing how creative thinking goes together with research and innovation in science. He didn't see it as a separate activity." Gimzewski added that "we both have the same interests. . . . [We don't want] to be confined in a field. My work is between areas. I've always been interested in doing art; Victoria was interested in science and math; [like me, she] went into a new area." "At the core," said Vesna, "we share incredible curiosity and wanting to discover something new. We're willing to take big risks."[20]

Vesna and Gimzewski worked well with Johnston and Lee, who shared a project-based approach to work. Vesna characterized the challenge of working with Hayles as of a clash of disciplinary cultures:

> I found it perhaps the most challenging and difficult to work with the literature group because of the difference of the process. . . . [T]he process is radically different. Radically. More than I ever thought. Whereas the similarity [between] science and media arts is usually that [we both work in] groups, are goal-driven, deadline-driven, and funding-driven. So there's such similarity there that it's very easy to connect and . . . relate. And there's a respect for whoever is the conceptual leader to have the goal and just go with it. With literature, it's very individual-oriented. . . . [O]ne person . . . figures things out, and [they] usually [do not] work in teams; it's not related to funding in any way, it's not technologically . . . driven, where you need others to help you realize something.

Gimzewski, Vesna and their students share the work model of the project-oriented science laboratory, in which students find dissertation topics in the course of working on large problem domains defined by the laboratory director. Research problems are often shaped by grants from organizations such as the National Science Foundation, which set deadlines for reporting progress and demonstrating results. Vesna viewed Hayles' disciplinary culture as "individual-oriented." From Vesna's perspective, this culture of individual work created friction with the project-oriented, team culture Vesna and Gimzewski shared with Johnston and Lee, and hindered the holistic dynamic she sought to foster among team members.

As Vesna and Gimzewski moved forward with their vision of the project and their students began working out the details of how to implement this vision, four of the other principals, Hayles, Sain, Eliel and Williamson, perceived themselves as relegated to advisory positions. The architects, by contrast, managed to remain central without appearing to assert themselves because all of the principals respected and relied upon their expertise. Johnston also that observed the culture of architectural practice traditionally looks to the client to define the design problem; part of the challenge for them was to negotiate a design solution acceptable to both Sain and Vesna, because both saw themselves as the primary client. Johnston, like Gimzewski, found that she and Lee were challenged by Vesna and Gimzewski to work in a different way. Instead of moving linearly toward a single solution, the architects engaged in a dialectical creative dialogue with Vesna and Gimzewski, generating ideas in conversation with the artist and scientist's evolving concepts for *nano*. Johnston later reflected that the project had influence their own perception of disciplinary boundaries as she and Lee began to work more in the manner of artists.

Hayles viewed the difficulties of bridging differences between disciplines not in terms of differing work cultures, but in terms of the inadequacy of her ideal of collaboration. Long before *nano*, Vesna and Gimzewski had established the kind of intellectual marriage of minds Hayles considered ideal. From Hayles' perspective, the marital metaphor for collaboration faltered when she, Sain, and Eliel challenged Vesna and Gimzewski to open up their dyad to include others as autonomous co-equals. In Gimzewski's organic metaphor, by contrast, individuals and disciplinary structures were subsumed in the symbiotic relations between types of permeable cells that make up living tissue.

When Collaboration Isn't Ideal: Part One

Rents in the collaborative fabric of the *nano* project appeared quite soon. The first significant collaborative crisis among project principals developed during the first months of meetings. Tension over how to work together mounted between Sain and Eliel on the one hand, and Vesna and Gimzewski on the other, with Johnston and Lee caught in the middle. At the same time, Hayles began to feel shut out of the creative process. Johnston and Lee gradually reduced their participation in meetings to spend more time working on the project in their studio. Vesna remarked that the LACMALab people did not seem to trust her and Gimzewski. Sain characterized their mutual frustration as a "normal reaction" to any process of creating an exhibition. Yet while he realized that this process would somehow be different, and even welcomed the experimental aspect of this collaboration as part of LACMALab's mandate, he persisted in his conviction that Vesna should respect his experience creating previous exhibitions at the Boone Gallery. "We have to see the big picture," said Sain:

> We're sitting with the artist [thinking about] ... what happens when a two-year-old walks in. What's the dynamic for the two-year-old, what's the dynamic for the parent? What happens when the two-year-old comes in with a grandparent? What happens when teenagers come in? What happens when you have every age under the sun in there?

Sain's claims can be construed as an attempt to assert epistemological authority over the territory of the museum. Sain argued that "we are the advocate for the audience. We've watched half a million kids go through artists' installations. ... We can tell you how kids are going to react." In addition, he saw himself as the expert in the material limitations and requirements of the Boone Gallery space: "We can see things like safety and accessibility. We expect artists to respect our expertise in that area."

Again and again in meetings, discussion of process eclipsed discussion of content because of disagreements among the principals about creative control. This dispute can be seen as a question of jurisdiction over museum space. This contested space became a field on which a deeper controversy over the boundaries of art and the museum played out. The museum can be seen as the space in which artistic concept becomes material object. In the creation of "Seeing" and "Made in California NOW," LACMALab's past shows, Sain had blurred the boundaries between the territories of artist and gallery director. He had imagined that *nano* would lead to a further blurring of this border. Although Vesna and Gimzewski worked to break down the boundaries between art and science, as well as between individual subjectivities, in their dialogue with the museum, they defended the boundaries of the domain of artistic practice: gallery space could not be separated from art installation.

As the weeks and months passed, and the date set for formal presentation of the exhibition proposal to the entire LACMA curatorial staff grew nearer, Sain began to realize that the collective, egalitarian ideals of collaborative process he thought had been agreed upon were not materializing. He came to believe that Vesna had a very different notion of collaboration. "We all experienced" that Victoria saw herself as the artist with others helping to carry out her vision. "That's where there was a disconnect" between how the process was described in meetings at the outset and how the process actually unfolded. "Part of the conundrum," Sain observed, "was ... that this was based on a preexisting relationship, preexisting project, and preexisting work... which we agreed to, we acknowledged. This was the only way that this was going to happen. But [we] did not think this preexisting situation would preclude a totally more collaborative way of working. ... We never saw this as a one-person show."

At the same time, Eliel also began to perceive that Vesna saw herself as the artist and LACMA chiefly as an exhibition venue. "[W]hile [Vesna] understood the importance of LACMA as an institution that would showcase her work," said Eliel, the curator did not have the sense that Vesna and Gimzewski considered the other principals as co-equals in creating the exhibition:

[I]t became fairly clear fairly quickly to Bob and to me that Victoria and Jim really were not interested in creating an exhibition that way. . . . They really were working on their own, arriving at what they wanted to be in the exhibition, and they would come back and present to the group, not as a possibility or as "this is a concept that we could chew over and flesh out and then perhaps change or perhaps discard," but that they were coming to present, "this is the next step," . . . and that was a very uncomfortable situation. . . . I think that they perceived themselves as the creative engines of the project, and the notion that Williamson or Kate or Bob or I could have . . . substantive creative input into the exhibition–really I don't think that they thought much of that concept.

During this phase, when tensions between principals were spiking, Johnston and Lee believed that their part of the assignment–designing a physical infrastructure of the exhibition that would mesh seamlessly with the art installations–helped defuse the crisis by moving the group's energy away from process and back to content. Rather than participate verbally in disagreements that arose during meetings, they concentrated on listening, then returned with drawings and models. "I think the objects should speak for themselves," Johnston said. "We usually present more than one, which I think helps to engage [people] in a dialogue." These material objects helped give synthetic, embodied form to the group's diverse ideas. The objects, Lee told me, communicated the architects' point of view non-verbally. "We had a . . . certain vision about the project," said Lee. "It's important that the models and the diagrams become 'Trojan horses' for us."

Early on, Johnston and Lee had found common ground with Vesna, Gimzewski and Hayles in their mutual enthusiasm for Buckminster Fuller's work and thought, which had implications for a number of disciplines, from chemistry to architecture. The architects were able to show how concepts drawn from Fuller's thought could be used to fashion a powerful yet flexible matrix for the evolving vision of the exhibition. The models gave everyone something concrete on which to focus, Lee observed, something in which all could see their ideas; this, he believed, helped everyone move forward. By constructing what I would call "transdisciplinary objects," Johnston and Lee made architecture function as a sort of fluid, neutral border zone where art, science, literature, and the museum could commingle.

While the drawings and models helped shift the group's focus from process to content, Hayles thought that the resolution of the crisis over the nature of the collaboration occurred when the presentation of the exhibition proposal met with enthusiastic approbation from Sain's colleagues at LACMA. For the first time, LACMA allocated significant advertising money for a Boone Gallery exhibition. It was clear that LACMA thought this was going to be an important show. This imprimatur of mainstream curatorial authority reassured Sain. He said it was at this point that, with realities of time and budget closing in, he let go of issues of process:

You're dealing with such sort of ingrained, basic questions about how people work together. . . . This stuff takes time. . . . If we had another year, in terms of how you build these expectations and relationships and clarity, it would probably be a different situation. . . . [W]hen it's all said and done, you can have all the aspirations in the world, and hopes and desires of shifting ways of working in the world, but you reach a moment that is just a plain business reality... You sit there and look at the calendar and [think] "Yikes! We have a show to open now;" we're past the stage of some of the more conceptual processes [of] re-imagining and re-envisioning.

Sain relinquished his hopes about pioneering a new collective way of working, and changed his conception of his role from equal collaborator to facilitator:

You would see the work, you would be aware of the reality and then you'd think, the smartest thing we can do at this point ... [is] to create a situation where [Vesna and Gimzewski] can do their best. It's not going to be of ... value if we get so stuck on certain things about process and about intent that the public is not necessarily going to realize.

Eliel, too, was reassured about the show's quality, even though she remained disappointed about the process:

I think we're both very pleased and excited about the show ... [S]o it's not that we feel, necessarily, badly about the product, but the process that we had hoped for and envisioned simply wasn't the process that led to the product.

While Sain and Eliel saw the process as less than ideal, ceding a significant portion of their authority over the museum space to Vesna and Gimzewski, Vesna appeared conflicted between the desire to adhere to her ideal of complete collaborative equality, and a need to assert authority as an artist. At times she viewed all of the team members as equal collaborators, including the students. At others, she spoke of herself and Gimzewski as co-equals leading the collaboration together. At still other moments, she acknowledged that Gimzewski deferred to her expertise as an artist. In a reflective moment, Vesna talked frankly about this conflict:

You find yourself in the center of negotiating how [the exhibition] maintains [the] conceptual coherence you originally envisioned while you still allow for that dialogue to happen. How you keep an aesthetic integrity so it doesn't become a mish-mash. Where you draw the line and where you allow it to open up. It really is a challenge for the artist, and for the museum as well.

Vesna's view of this tension as boundary work resonates with the notion of the museum as a zone where artistic practice and curatorial practice negotiate their borders. Among the UCLA principals, the balancing act that Vesna described also reflects the professional expectations of different academic disciplines, which dictate that Vesna would, in the end, bear the greatest share of accountability for the success or failure of what was, after all, an art exhibition. Hayles and Gimzewski's reputations rested on their achievements in literature and in nanoscience. Clearly they had much less professionally at stake in *nano*'s success than Vesna, an artist. On a purely practical level, Vesna's conflict reflects the project's fundamental need for creative leadership—a need that the art and science students, as well as the architects, saw as a common sense given. From their perspective, Vesna and Gimzewski were the natural and unquestioned leaders of the project.

Collaboration and Power Relations

The moment Vesna faced her previously unacknowledged conviction that the principals were not co-equal collaborators came when the project's principals faced their second major collaborative crisis, the departure of Williamson.[22] Williamson, a freelance writer without the stature, salary, or resources that come with an institutional position, told me in an informal exchange that she felt increasingly caught between, and ultimately marginalized by, two large and powerful institutions: UCLA and LACMA; according to Hayles and Sain, she articulated the same feelings to them more than once. While Sain and Hayles became resigned to being pushed out of collaboration on the conceptualization of the exhibition, Williamson became increasingly uncomfortable with her position and frustrated at having her ideas ignored.[23]

Her frustration reached boiling point during a meeting between the Vesna, Gimzewski, Sain, Eliel, Williamson and the head of LACMA's South and Southeast Asian collection, Stephen Markel. A group of Tibetan monks had been asked to create a sand mandala at LACMA during a fall 2003 exhibition on Buddhist art, "Circle of Bliss," at LACMA's main site, LACMA East. Upon learning this, Gimzewski and Vesna became interested in the possibility of making a connection between the creation of the sand mandala and nanotechnology. Sain, Eliel and Markel liked the idea of using the mandala in *nano,* not least because it would link the exhibition in LACMA East with *nano* in LACMA West, site of the Boone Gallery. Vesna recalled that Markel spoke of "the building of the sand mandala as the first nanotechnology, as the Tibetans [had] used, in the past, ground minerals and gems to [symbolically] build an entire universe from a single speck."[24] When approached, the monks, too, saw the connection and responded favorably to the proposal.

Williamson, by contrast, objected to the idea. Much of Williamson's previous work focused on analyzing and critiquing the shaping presence of spiritual and religious notions in science and technology. From Eliel's perspective, Vesna and Gimzewski were not willing to open their minds to Williamson's point of view:

> I felt strongly that in this particular meeting, Victoria and Jim had a fixed idea ... [T]hey knew what they wanted, and there was no way they were going to leave not having that. They just were not going to listen to what Williamson had to say. And I think ultimately, that's when Williamson decided, "I don't need this in my life." ... There was an unwillingness to listen to her on the part of Victoria and Jim. ...I actually happened, philosophically, in this particular instance, to side with ... Victoria and Jim because I did feel it was appropriate to include [the mandala]. But I thought it would have been useful to have the full conversation, to understand further what Williamson's concerns [were].

Hayles did not attend this meeting, but when she heard about the sand mandala proposal, she agreed with Williamson. Like the science writer, Hayles had deep reservations about the appropriation of eastern philosophy by western scientists. Hayles and Williamson joined in strongly opposing bringing the Tibetan Buddhist sand mandala into *nano.* With Sain, Eliel, and Markel assenting, however, Vesna and Gimzewski went ahead with their plans.

A few weeks later, after long conversations with Sain about her disillusionment with the collaborative process, Williamson resigned from the project—just as the chief portion of her contribution, the selection and treatment of text in the exhibition, was about to begin. Her departure came as a shock to Hayles, Vesna, and myself. Vesna expressed regret at Williamson's departure, and tried to contact her to discuss it, but Williamson would not see her. Looking back at this moment several months after Williamson withdrew, Vesna and Gimzewski faced the reality of their heretofore unacknowledged belief that the other principals did not have equal voices in the conceptualization of the exhibition:

> Jim and I talked about it and we admitted to ourselves [that] if Jim and I disagreed strongly with something, it would not happen. If Williamson refused to have monks in there, we would say "tough," because ultimately there is something to be said about who takes full responsibility and commitment for making the project happen. . . . It's who came up with the concept initially, [who] came up with funding and people, who is committed full-time to working on this.

Vesna argued that "If [Williamson] . . . made the choice to [invest] an equal amount of commitment and time and energy into this, she *would* be equal." Vesna's statements about the possibility of equal collaboration among principals highlight a significant tension between egalitarian ideals about working together, and the project's practical need for creative leadership. Here, the claim to leadership rested on differential abilities to marshal resources for the project. On another level, Vesna framed the dispute over the sand mandala in terms of shifting disciplinary boundaries. She pointed out that profound suspicion toward religion and spirituality pervades much postmodernist theory. Postmodernism, which exerted a strong influence on literary and cultural studies in the 1990s, had, like Western science, more or less ruled religion and spirituality out of disciplinary bounds; the resistance of Hayles and Williamson to the sand mandala illustrated the lingering influence of postmodernism on their thinking. Vesna and Gimzewski's work together, by contrast, had begun to challenge this dogma. They were open to exploring ways in which the boundaries of science and art could expand to incorporate spirituality.

Collaboration and Conflicts Over Textual Turf

The third crisis among the principals occurred over the issue of the selection of text to be used in the exhibition. Vesna explained the conflict in terms of clashing work cultures. For Hayles, it was a matter of respect for the expertise that long disciplinary training bestowed. Viewed in terms of boundary work, Hayles, having felt excluded from the conceptualization of the exhibition, staunchly defended the disciplinary territory of literary studies, while Vesna sought to incorporate text into the realm of art.

LACMA also felt a vested interest in the text. For Sain, being in control of text selection was normal procedure: LACMA had always presided over the choice and appearance of text in their galleries. The difference in *nano* was that text was not just explanatory or didactic, unrelated conceptually to installations; here, text had an important relationship to concept. This complicated the issue of disciplinary jurisdiction.

The process of text selection began with a list of themes and motifs related to nanotechnology. Professors and students joined to brainstorm the list at the January "synaptic blow-out" event. Hayles typed the ideas up and distributed copies to the entire team. In the spring, Hayles asked the literature students to contribute as many quotations from as wide a range of sources as possible, using the list of themes and motifs as a guide. In my capacity as Hayles' research assistant, I sent an e-mail to the rest of the team as well, inviting contributions. I uploaded first the list of themes and motifs, and the compiled list of passages to the group's collaborative website, letting everyone know it was on the site. Hard copies of the compilation were widely distributed at meetings.

During the spring of 2003, Hayles, Sain and Eliel met at LACMA to hammer out a conceptual framework for text selection. Hayles based her ideas for the conceptual framework on a thematic brainstorming document produced by the entire UCLA team. Sain's style in these meetings was to pose questions, sometimes in the form of divergent alternatives, and then allow the group to hash out answers acceptable to all. Hayles then chose passages according to these guidelines. Hayles said that she felt her meetings with Sain and Eliel were a satisfying and productive phase of the project for her. Vesna, however, later expressed frustration at being excluded from these meetings. Like the gallery space and the architectural drawings and models, text began to function as a border zone situated on the contested frontier between art, science, literary studies, and the museum.

When Hayles, Sain and Eliel had agreed upon the framework, it was circulated to the entire group and discussed at a retreat in mid June of 2003. At the retreat, spirited debate arose over two passages: one drawn from Eric Drexler's popular work of futurist speculation, *Engines of Creation,* and one from Michael Crichton's novel *Prey.* The argument highlighted a difference in how text's function in the exhibition was viewed, a difference clearly arising from divergent disciplinary frameworks. Gimzewski and his students objected to the inclusion of anything by Drexler or Crichton. The representations of nanoscience in these texts were simply wrong, Gimzewski said, and Vesna concurred. Hayles and several of the literature students, including myself, argued that the issue was not the passages' factuality; the exhibition's purpose was not didactic. Rather the issue was to represent how certain texts have shaped popular beliefs about nanotechnology.

The disciplinary boundaries were sharply defined in this disagreement: From the perspective of Hayles and the literature students, Gimzewski and some of his students appeared to view the main purpose of text as the transmission of factual information. Hayles and her students emphasized the mutually constitutive relationship between language and scientific discovery; and the cultural influence of text—in this case, the various ways that texts circulating throughout the culture constitute nanotechnology as a discourse.[25] Hayles offered a compromise: since Gimzewski objected so strongly to Drexler, she would excise Drexler from the passages. An excerpt from *Prey* would remain, as creative fiction, rather than accurate description. She again invited everyone to contribute any other passages they felt were improvements on those she had selected.

The disagreement and its resolution at the retreat suggests that Hayles remained open to negotiation over specific passages as long as her broader disciplinary jurisdiction over text as the purview of literary studies was not challenged. The issue of text selection arose again, however, at a team meeting in the EDA at the beginning of July. Since classes had ended for the year, most of the literature students were present, as were many of the art and science students. In the course of the meeting, Vesna said that she wanted some of Gimzewski's students to present passages they had selected and desired to contribute.[26] Hayles interrupted Vesna to say that all of the passages had been carefully selected to fit the conceptual framework that LACMA had approved. All of the passages, she argued, existed in an overall "balance" with one another; passages could not be considered outside this framework. Vesna replied that while the literature contingent was out of touch with the exhibition's design, the science students were in the thick of the development process; therefore, she said, "we feel it's important to hear what the science students came up with." She proposed that the entire group revisit the original compilation of passages and discuss the selection of passages together. A dispute ensued between Hayles and the literature students on one side, and Vesna and the art and science students on the other. The crux of the issue was that Vesna and some students had become increasingly concerned that they were being excluded from the process of choosing text and determining how selected passages would be used. From their perspective, Hayles was not collaborating with them. Hayles insisted that she was collaborating—with LACMA. According to Vesna, Hayles said there were actually two collaborations occurring: one between the science and art groups, and one between the literature and LACMA groups.

The public phase of the argument ended only through mutual exhaustion. I and some of the other literature students lingered outside the building afterwards, discussing what had occurred. We felt angry and defensive, on behalf of ourselves as well as Hayles.

Later I learned from Hayles that the dispute between herself and Vesna continued in private, through a series of telephone and e-mail exchanges. Hayles at this point demanded that the final decisions about the text be hers.

Vesna more or less acquiesced, but did not accept the implicit division of labor. Some months later, she invoked the transdisciplinary concept of holism to argue that the conceptual integrity of the exhibition was at stake: "Actually, it is about thinking in a holistic manner, and as soon as that becomes the mode of exchange, [the process becomes] a flow, and it is much easier for everyone involved to come to a place of comfort [with decisions]."[27] From her perspective, the text "was treated as [being] separate from the installations" and did not "reflect the thinking behind the projections, i.e., the buckyballs, the hexagonal floors, the molecular shapes, the Fuller [design principles], . . . as the architecture [of the exhibition] did."

In an elaboration of her argument about holism, Vesna later suggested that the limited participation of Hayles and her students in the work of creating the exhibition exemplified the historical division of theory from practice in the academy, such as happened in the arts "when art history departments separated from art departments and moved to the humanities." Thus the problem, she asserted, stemmed from "separation": "I believe that that the literature group separated," and that this "created problems and miscommunications."[28]

Vesna's claim about holism as a principle to which the rest of the team should have submitted exemplifies the way that transdisciplinary rhetoric can simultaneously mask and shape power relations.[29] What is more, Vesna's argument about theory versus practice assumes that making art is a material practice, while writing is not. In contrast, Hayles viewed the substantial commitment of time and energy that her students made to writing essays for this book as fully equivalent to physical work.[30]

It was finally on the playing field of power that Hayles chose to negotiate this boundary dispute with Vesna. Hayles told Vesna she was willing to resign from the project and let her choose all of the text. Hayles had several valuable chips on her side of the table: this book, for which she had secured publication, and the symposium to be held in conjunction with the exhibition, which she was to organize.[31] Beyond this, there was Hayles' professional stature, which brought to the exhibition her widely respected expertise as a theorist of science and culture. Finally, Vesna decided not to force the issue, and told Hayles she wanted her to remain in the project.

What exactly happened here? Upon reflection, it seems clear that differing views of the nature of the collaborative process played a role in fomenting the dispute between Hayles and Vesna over the territory of text. Vesna and Gimzewski, as seemed natural to their students, acted as the creative heads of the project. Vesna's idea of collaboration was based on a project model with a clear leader, such as a lab director, film director, or senior architect. She solicited ideas and labor from team members, gave them assignments, and allowed them considerable creative scope, but, Vesna, along with Gimzewski, made the final creative decisions.

Hayles had abandoned the ideal of egalitarian collaboration as far as Vesna and Gimzewski were concerned. In her work with the LACMA staff, however, Hayles experienced a highly satisfying and productive dialogue among equals. Hayles respected the experience and training that the LACMA staff had with staging large-scale, participatory exhibitions for audiences of all ages; the LACMA staff respected Hayles' expertise in literature and science studies. Each respected the boundaries of the other's disciplines. Eliel contrasted her work with Hayles on the text with her experience with Vesna and Gimzewski: "We [Hayles, Eliel, and Sain] didn't always agree; [Hayles] wanted things that we didn't want; we wanted things that she didn't want. But we always were able to have an intelligent discussion about it, and arrive at a place [where] everybody felt comfortable with the solution." Frustrated in her attempt to collaborate with Gimzewski and Vesna in the conceptualization phase, Hayles staked out a piece of disciplinary turf and defended it. This defense involved a shift in Hayles' model of collaboration on the project from equal creative partnership between all eight principals to a model of two groups working separately on different aspects of the exhibition: the science and art groups, and the literature and LACMA groups.[32]

Looking back a few months afterward, Hayles articulated the nature of her disciplinary terrain, and why she had resisted Vesna's attempt to extend the boundaries of art and science to encompass text. Hayles characterized it as a divide between areas of disciplinary expertise:

> The point at which [gaps between disciplines] became obvious to me had to do with other people's understanding of what literature is, and of what text is. I'm sure I presented comparable blindnesses of which I was not aware that other people in other disciplines could see, so I don't mean to privilege these, it's just that because of my disciplinary training, I happen to be able to see this with particular clarity. ... So, this is not to try to find fault in any way; it's just simply to recognize that this is an area of disciplinary knowledge that has real effects, and if you know some things your prac-

tices are different than if you don't know those things. . . . [W]hat is necessary is a deep understanding and practice in the very specific micro-effects that language has.

Vesna, on the other hand, framed the dispute in terms not of disciplinary expertise, but of holism and differing work cultures within disciplines:

There is [at] the core of it a misunderstanding of how we work, and what the creative process is. . . [T]he biggest challenge to me came when we would have these endless discussions in media arts and science trying to figure out how things would work. And then we would just be told, "OK, here's the text." That . . . didn't fit, because the fundamental process is to make it all mesh . . . into a kind of hybrid. So you can't just go into your room and figure it out and come back and do a report. It just doesn't work. On the literature side I think—and again this is just me guessing—I think there was a sense of like "god, you know, we did all this work, put in all this effort, and then we come back, and they're acting . . . aggressive and not understanding how much time and energy went into this," right? And on this side [we're saying], "wait a second, we're doing all this work, we're putting it all together as a group and they just come in and just hand us stuff that has nothing to do with what we were talking about."

Vesna's explanation of the dispute stemmed from a commitment to transdisciplinarity, and an attempt to enact this theoretical stance as a different way of working while simultaneously maintaining creative authority. Hayles' view, by contrast, reflected a commitment to disciplinary grounding as a way to achieve expertise. In part, institutionally enforced disciplinary structures blocked efforts to fashion a holistic work process, while long enculturation into academic disciplinarity proved inadequate in the context of a transdisciplinary, project-based collaboration. Entrenched in institutional power relations, neither scholar could find the ideal formula for working together.

Conflict and Creativity

In general, despite multiple disputes and the departure of one of the principals mid-project, most of the main players embraced conflict as a creative and transformative dynamic, and emerged well-satisfied with the exhibition. Looking back in early October of 2003, James Gimzewski pronounced himself "very, very happy" with both the process and the product: "[I]n this kind of constant motion of conflicts, of happiness, of tears, in this comes this marvelous creation" that he could not have foreseen. It was "just an incredible learning experience, an incredible experience for my students, the art students, for the literature people. . . . The actual show itself is one thing, but the actual experience, the path, going along the path to get to the show, to me is worth it, for that alone".[33]

"Conflict is part of the process," he reflected. "The problem is how you deal with conflict. If you have a conflict, and in that conflict you learn and change and produce something from that conflict, [it] can be something very positive, very new."

Project principals ultimately came to a provisional understanding that enabled the work to continue. Like Gimzewski, Vesna viewed conflict positively. "Gradually," she said:

There was more of an understanding on both sides, because ultimately, I think, every-body wants it to work. And everybody sees the potential of these disparate groups coming together toward one goal, which is to create something different: another way of showing how science and art and culture intersect, right? So it's resolved just by tenacity, I think . . . that's almost irrational. . . . because it doesn't make any practi-cal sense. I think there's just a kind of deep, intuitive belief that this is the way to go. And it's so challenging but it's interesting, you know. It's so much more interesting to be in this atmosphere that brings up emotions and passions and interest and ideas than in an environment where everything is predictable and set, and you know what your path is. I think ultimately everybody just hung in there.

On her part, Hayles declared herself "very happy" with the exhibition: "There was a lot of give-and-take, . . . But finally we were able to work that out, . . . and I'm very happy with the exhibit as a whole."

Why did most of the principals persevere despite moments of profound misunderstanding, frus-tration, anger, and alienation? In the final analysis, their persistence can only be explained by a deep, common commitment to seeing the project succeed, as well as to a belief that attempts to initiate cross-disciplinary dialogues matter, regardless of boundary disputes.

Collaboration's Future

Perhaps the real collaborative success story lies among the students, some of whom engaged in exciting and inspiring dialogues across disciplines, and, particularly Vesna's and Gimzewski's students, worked closely together for many months. A number of art and science students, as well as some of the literature students, immediately found common ground during their dialogues at the "synaptic blow-out." They came away stimulated and inspired by contact with other disciplin-ary perspectives. Some of these conversations grew into highly productive work relationships and friendships.

A literature student identified the first event as a key moment for interdisciplinary dialogue. Students were sent off in groups to engage in free-ranging, open-ended discussions about their interests and how these interests might intersect. He describes the excitement he felt about the potential for collaboration to blur disciplinary boundaries:

> I definitely think the most exciting moment of the process for me was probably the first meeting at Kate Hayles' house, when we all got together, and just started talk-ing things over in an informal way. And right away it was really engaging just to be involved with people from another discipline, to get a perspective completely outside what I'm used to. And I think there was a lot of excitement generated that night about projects that were thought [of] right from the start as a kind of indeterminate col-laboration. There wasn't really an agenda set. And that's what I felt was really excit-ing about it, that it felt more as though the students were sitting down together and determining the process. . . .The thing that was exciting for me that night in particular was that I have an interest in visual poetry, kind of avant garde poetic practices, and

right away that night I was able to talk to several of the science students, and to [the student] in communications, and they seemed really interested in what I was saying. A lot of the projects that they were working on, for example [a science student's] project with the AFM [Atomic Force Microscope] and the production of sounds from cells with the AFM was really interesting to me in relation to my interest in visual poetry and interfaces and electronic music and things like that. Conceptually I think there really was a real connection. . . . all four of us [in our group] were able to exchange ideas in a really effective way.

As the project neared its opening date, one of Gimzewski's students who worked closely with a media arts student on sound spoke of the transformative impact of their experiences collaborating on *nano:*

I have developed a close working relationship with [one of the art students]. . . and now we are submitting a video art proposal to the Hammer Museum, and we are discussing a proposal for the Banff New Media Institute. Most importantly, it is our collaboration which will be the basis for her MFA thesis and a large section of my Ph.D. thesis. I hope to continue to work with her and [with] many other artists I have met thus far. . . . *nano* has changed the way I see myself, my research and my goals in life. . . . I am changed, and I am the better for it.

For Vesna, seeing these connections grow among students brought her tremendous validation as a teacher:

There [are] moments [when] you realize it's absolutely worth it, and it can be so satisfying. And I think those moments are when I walk into the [Pico] lab and I see two of my students there . . . passionately engaged in a discussion with some of Jim's students. [N]obody arranged this; they just completely got together [on their own], and they have this need to figure things out. [I]t's just fantastic. You walk in and you feel like "this is great, and it's really worth it." . . . Or if I see [one of the literature students] describing the STM [Scanning Tunneling Microscope] to the whole group after just a couple of meetings I feel, "yeah, this is what it should be about." . . . [T]his is our responsibility in this academic setting, if we have the positions we hold, not just to sit there on it, but to take those kinds of risks where we create the space for experimentation, and for students to look at things from a different perspective.

Students' experiences finding common conceptual ground on the *nano* project—especially Vesna's and Gimzewski's students—suggest that the most promising potential for building bridges between disciplines lies among those who have not yet secured powerful institutional positions, or hardened their disciplinary boundaries through long training. For more professionalized practitioners, high walls must often be breached before bridge-building can commence. Only through sustained commitment to long-term dialogues that establish communicative competence, trust, and respect can interdisciplinary collaboration succeed.

On *nano,* boundary work and collaboration proved to be inextricably intertwined. Boundaries between disciplines did not manifest themselves as static or fixed perimeters. As disciplinary cultures came into contact through collaboration, fluid, shifting border regions arose that manifested at different moments as a physical space—the museum gallery—or as a material object, such as an architectural model, or as passage from a popular novel. Closer attention to dynamics in liminal zones such as the ones I have described can help those committed to interdisciplinary and transdisciplinary work understand what happens to epistemic boundaries at the limits of disciplinarity.

○○○

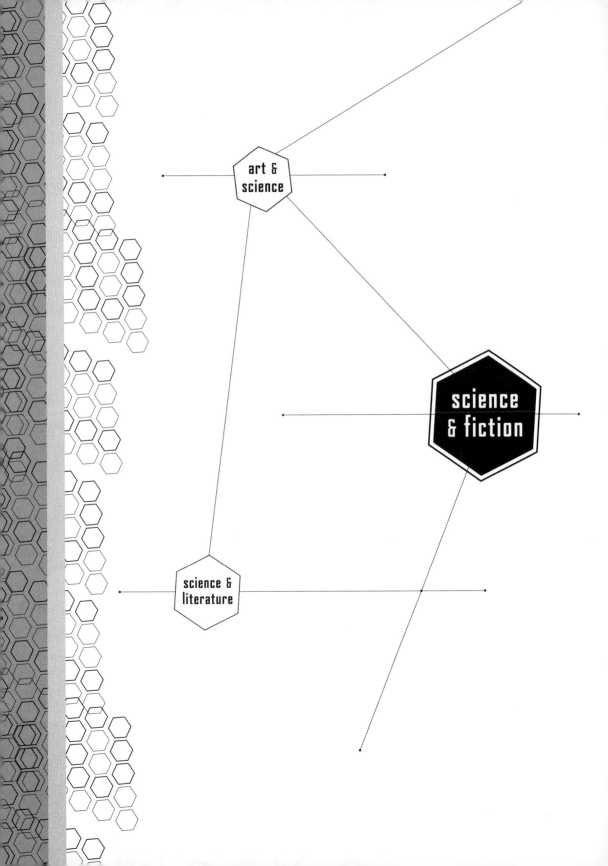

art &
science

science
& fiction

science &
literature

Nanotechnology in the Age of Posthuman Engineering: Science Fiction as Science

COLIN MILBURN

> Now nanotechnology had made nearly anything possible, and so the cultural role in deciding what should be done with it had become far more important than imagining what could be done with it."[1]
>
> ◎ Neal Stephenson, *The Diamond Age* (1995)

> Long live the new flesh.
>
> ◎ David Cronenberg's *Videodrome* (1983)

The Technoscapes and Dreamscapes of Nanotechnology

K. Eric Drexler, pioneer and popularizer of the emerging science of nanotechnology, has summarized the ultimate goal of his field as "thorough and inexpensive control of the structure of matter."[1] Nanotechnology is the practical manipulation of atoms; it is engineering conducted on the molecular scale. Many scientists involved in this ambitious program envision building nanoscopic machines, often called "assemblers" or "nanobots," that will be used to construct objects on an atom-by-atom basis. Modeled largely after biological "machines" like enzymes, ribosomes, and mitochondria—even the cell—these nanomachines will have specific purposes such as binding two chemical elements together or taking certain compounds apart, and will also be designed to replicate themselves so that the speed and scale of molecular manufacturing may be increased. Several different types of assemblers, or assemblers with multiple functions, will act together to engineer complex objects precise and reproducible down to every atomic variable. With its bold scheme to completely dominate materiality itself, nanotechnology has been prophesied to accomplish almost anything called for by human desires.

These prophesies have run the gamut from the mundane to the fantastic. Nanomachines will be able to disassemble any organic compound, such as wood, oil, or sewage, then restructure the constituent carbon atoms into diamond crystals of predetermined size and shape for numerous purposes, including structural materials of unprecedented strength. Nanomachines will be put into your carpet or clothing, programmed to constantly vaporize any dirt particles they encounter, keeping your house or your wardrobe perpetually clean. Nanomachines will quickly and cheaply fabricate furniture, or car engines, or nutritious food, from a soup of appropriate elements. Nanomachines will facilitate our exploration of space, synthesizing weightless lightsails to propel seamless spaceships throughout the universe. Nanomachines will repair damaged human cells on the molecular level, thus healing injury, curing disease, prolonging life, or perhaps annihilating death altogether.

Nanotechnology has been extensively discussed in these terms, but despite the fancifulness of certain nanoscenarios, it has become a robust and active science. Many universities, laboratories and companies around the world are investigating nanotech possibilities, constituting a dense discourse network—a technoscape—of individuals and institutions interested in the potential benefits of this nascent discipline.[2] The U.S. National Science Foundation supports a National Nanofabrication Users Network to coordinate efforts at numerous sites,[3] and the National Nanotechnology Initiative, proposed by the Clinton administration in 2000 and augmented by the Bush administration in 2001, offers funding and guidelines to promote nanotech breakthroughs.[4] Arguably at the center of the technoscape is the Foresight Institute, a non-profit organization established in 1986 by Drexler and his wife, Christine Peterson, to foster thinking and research related to nanotechnology.[5] Hosting conferences, sponsoring publications and awards, the Foresight Institute strives to be a nanotech mecca of sorts, anchoring the morass of nanotechnological endeavors currently spreading across the globe. Since Drexler first proposed a program for research in 1986 with the publication of his polemical *Engines of Creation: The Coming Era of Nanotechnology,*[6] nanotechnology has gained notoriety as a visionary science and the technoscape has burgeoned.

Offering intellectual and commercial attractions, career opportunities and research agendas, nanotechnology foresees a technocultural revolution that will, in a very short time, profoundly alter human life as we know it. The ability to perform molecular surgery on our bodies and our environment will have irrevocable social, economic and epistemological effects; our relation to the world will change so utterly that even what it means to be human will seriously be challenged. But despite expanding interest in nanotech, despite proliferating ranks of researchers, despite international academic conferences, numerous doctoral dissertations and hundreds of publications, the promise of a world violently restructured by nanotechnology has yet to become reality.

Scientific journal articles reporting experimental achievements in nanotech, or reviewing the field, frequently speak of the technical advances still required for "the full potential of nanotechnology to be realized,"[7] of steps toward fulfilling the "dream of creating useful machines the size of a virus,"[8] of efforts that, if they "pan out, . . . could help researchers make everything from tiny pumps that release lifesaving drugs when needed to futuristic materials that heal themselves when damaged."[9] These texts—representative of the genre of popular and professional writing about nanotech that I will call "nanowriting"—incorporate individual experiments and accomplishments in nanoscience into a teleological narrative of "the evolution of nanotechnology,"[10] a progressivist account of a scientific field in which the climax, the "full potential," the "dream" of a nanotechnology capable of transforming garbage into gourmet meals and sending invisible surgeons through the bloodstream, is envisioned as *already inevitable.*

Nanowritings convey "a sense of inevitability that [future nanotech successes] will come in time," a sure faith that there "will come technologies that will be the best that they can ever be" and that "all manner of technologies will flow" from the current work of dedicated visionaries.[11] Because the "development of nanotechnology appears inevitable,"[12] nanowritings freely and ubiquitously import the nanofuture into the research of today, and the language used, as we will see, rewrites the advances of tomorrow into the present tense.[13] Nanowritings speculate on scientific and technological discoveries that have not yet occurred, but they nonetheless deploy such fictionalized events to describe and to encourage preparation for the wide-scale consequences of this "seemingly inevitable technological revolution."[14]

Even in the discipline's first recognized technical journal article—which both proposed a new technology and inaugurated a new professional field—Drexler writes that the incipient engineering science of molecular nanotechnology has dramatic "implications for the present" as well as the "the long-range future of humanity."[15] Repeated throughout the technoscape, this narrative telos of nanotechnology—described as already given—is a vision of the "long-range future of humanity" utterly transfigured by present scientific developments. In other words, embedded within nanowriting is the implicit assumption that, even though the nanodreams have not yet come to fruition, nanotechnology has already enacted the transformation of the world.

Due to the tendency of nanowriting to speculate on the far future and to prognosticate its role in the radical metamorphosis of human life (coupled with the fact that nanotech research has yet to produce material counterparts to its adventurous mathematical models and computer simulations) many critics have claimed that nanotechnology is less a science and more a science fiction. For instance, David E. H. Jones, chemist at the University of Newcastle upon Tyne, insinuates that nanotech is not a "realistic" science, and that, because its aspirations seem to violate certain natural limits of physics, "nanotechnology need not be taken seriously. It will remain just another exhibit in the freak-show that is the boundless-optimism school of technical forecasting."[16] Gary Stix, staff writer for *Scientific American* and persistent critic of nanotech, has compared Drexler's writings to the scientific romances of Jules Verne and H. G. Wells, suggesting that "real nanotechnology" is not to be found in these science fiction stories.[17] Furthermore, Stix maintains that nanowriting, a "subgenre of science fiction," damages the legitimacy of nanoscience in the public eye and that "[d]istinguishing between what's real and what's not" is essential for nanotech's prosperity.[18] Similarly, Stanford University biophysicist Steven M. Block has said that many nanoscientists, particularly Drexler and the "cult of futurists" involved with the Foresight Institute, have been too influenced by laughable science fiction expectations and have gotten ahead of themselves; he proposes that for "real science to proceed, nanotechnologists ought to distance themselves from the giggle factor."[19]

Several critics have stated that direct atomic manipulation and engineering is not physically possible for thermodynamic or quantum mechanical reasons; others have suggested that, without experimental verification to support its outrageous notions and imaginary miraculous devices, nanotechnology is not scientifically valid; many more have dismissed the long-range predictions made by nanowriting on the grounds that such speculation obscures the reality of present-day research and the appreciable accomplishments within the field. These attacks have in common a strategic use of the term "real science" opposed to "science fiction," and, whether rejecting the entire field as mere fantasy or attempting to extricate the scientific facts of nanotech from their science-fictional entanglements, charges of science-fictionality have repeatedly called the epistemological status of nanotechnology into question.[20]

Nanotechnology has responded to these attacks with various rhetorical strategies intended to distance its science from the negative associations of science fiction. However, I will be arguing that such strategies ultimately end up collapsing the distinction, reinforcing the science-fictional aspects of nanowriting at the same time as rescuing its scientific legitimacy. I hope to make clear that the scientific achievements of nanotechnology have been and will continue to be extraordinarily significant; but, without contradiction, nanotechnology is thoroughly science-fictional in imagining its own future, and the future of the world, as the product of scientific advances that have not yet occurred.

Science fiction, in Darko Suvin's formalist account of the genre, is identified by the narratological deployment of a '"novum"–a scientific or technological "cognitive innovation" as extrapolation or deviation from present-day realities–that becomes "'totalizing' in the sense that it [the novum] entails a change in the whole universe of the tale."[21] The diegesis of the science fiction story is an estranging "alternate reality logically necessitated by and proceeding from the narrative kernel of the novum."[22] Succinctly, science fiction assumes an element of transgression from contemporary scientific thought that in itself brings about the transformation of the world. It follows that nanowriting, positing the world turned upside down by the future advent of fully functional nanomachines, thereby falls into the domain of science fiction. Nanowriting performs radical ontological displacements within its texts and recreates the world atom by atom as a crucial component of its extrapolative scientific method; but by employing this method, nanowriting becomes a postmodern genre that draws from and contributes to the fabulations of science fiction.[23] Science fiction is not a layer than can be stripped from nanoscience without loss, for it is the exclusive domain in which mature nanotechnology currently exists; it forms the horizon orienting the trajectory of much nanoscale research; and any eventual appearance of practical molecular manufacturing–transforming the world at a still unknown point in the future through a tremendous materialization of the fantastic–would remain marked with the semiotic residue of the science-fictional novum. Accordingly, I suggest that molecular nanotechnology should be viewed as simultaneously a science and a science fiction.

Jean Baudrillard has frequently written on the relationship of science to science fiction, contextualizing the dynamics of this relationship within his notion of hyperreality. Mapping onto "three orders of simulacra"[24]–the counterfeit, the reproduction, and the simulation–three orders of the speculative imaginary are described in his essay, "Simulacra and Science Fiction." He writes, "To the first category [of simulacra] belongs the imagination of utopia. To the second corresponds science fiction, strictly speaking. To the third corresponds–is there an imaginary that might correspond to this order?"[25] The question is open because the third-order imaginary is still in the process of becoming and is as yet unnamed. But within this imaginary, the boundary between the real and its representation deteriorates, and Baudrillard writes that, in the postmodern moment, "There is no real, there is no imaginary except at a certain distance. What happens when this distance, including that between the real and imaginary, tends to abolish itself, to be reabsorbed on behalf of the model?"[26] The answer is the sedimentation of hyperreality, where the model becomes indistinguishable from the real, supplants the real, precedes the real, and finally is taken as more real than the real:

> The models no longer constitute either transcendence or projection, they no longer constitute the imaginary in relation to the real, they are themselves an anticipation of the real, and thus leave no room for any sort of fictional anticipation–they are immanent, and thus leave no room for any kind of imaginary transcendence. The field opened is that of simulation in the cybernetic sense, that is, of the manipulation of these models at every level (scenarios, the setting up of simulated situations, etc.) *but then nothing distinguishes this opera from the operation itself and the gestation of the real; there is no more fiction.*[27]

In the dichotomy of science versus science fiction, the advent of third-order simulacra or imaginaries announces that science and science fiction are no longer separable. The borderline between them is deconstructed. In Baudrillard's age of simulation, science and science fiction have become coterminous: "It is no longer possible to fabricate the unreal from the real, the imaginary from the givens of the real. The process will, rather, be the opposite: it will be to put decentered situations, models of simulation in place and to contrive to give them the feeling of the real, of the banal, of lived experience, to reinvent the real as fiction, precisely because it has disappeared from our life."[28] At the moment when science emerges from within science fiction and we can no longer tell the difference, the real has retreated and we are only left with the simulations of the hyperreal where "there is neither fiction nor reality anymore" and "science fiction in this sense is no longer anywhere, and it is everywhere."[29]

The case of nanotechnology illustrates the hyperreal disappearance of the divide between science and science fiction. The terminology of "real science" versus "science fiction" consistently used in the debates surrounding nanotech depends upon the discursive logic of the real versus the simulacrum as analyzed by Baudrillard. Although each term may independently provide the illusion of having a positive referent—that is, "real science" might refer to a set of research and writing practices that adhere to and/or reveal facts of nature while being institutionally recognized as doing so, and "science fiction" might refer to a set of certain generically-related fictional texts or writing practices that mimic such texts—when they are used to argue the cultural status of nanotechnology, real science and science fiction are nearly emptied of referential pretensions, becoming signifiers of unstable signifieds as they are forced into pre-established symbolic positions of "the real" and "the simulacrum." In this logic, science and science fiction negatively define each other, and though each is required for the other's structural existence, science fiction is the diminished and illegitimate term, the parasitical simulation of science.

To maintain that the categories of science and science fiction are supplemental constructs of each other is not to deny the political effects of discourse, for the fate of nanotechnology as a research field and the fates of real people working within it are strongly entwined with the language used. But I will show that the nanorhetoric mobilizing the logic of real science *opposed* to science fiction comes to undermine its own position, dissolving real science into science fiction and exemplifying what Baudrillard describes as the vanishing of the real, or the moment of hyperreal crisis when the real and "its" simulacrum are understood as semiotic fabrications, when "the real" (e.g. "real science") can be demonstrated as simulation and "the simulation" (e.g. "science fiction") can be demonstrated as real, when dichotomies must be abandoned in favor of hybrids. Although the strict categories of real science and science fiction must be used in order to accomplish their deconstruction (or are deconstructed because of their use), they should be read as under erasure, for the relationship of science to science fiction is not one of dichotomy but rather of imbrication and symbiosis. Science fiction infuses science and vice versa, and vectors of influence point both ways. Inhabiting the liminal space traversed by these vectors are fields like nanotechnology that draw equally from the inscription practices of scientific research and science fiction narration, and only a more sutured concept—something like "science (fiction)"—adequately represents the technoscape of nanotechnology and its impact on the human future.

Nanotechnology is one particular example illustrating the complex interface where science and science fiction bleed into one another. Yet more significantly, nanotechnology is capable of engineering the future in its own hybrid image. Not only does the continued development of nanotechnology seemingly provide the means for making our material environments into the stuff of our wildest dreams, but nanotech's narratives of the "already inevitable" nanofuture ask us *even now* to reevaluate the foundations of our lived human realities and our expectations for the shape of things to come. Which is to say that the writing of nanotechnology, as much as or even more than any of its eagerly anticipated technological inventions, is already forging our conceptions of tomorrow. Unleashing its science fictions as science and thereby redrawing the contours of technoculture, nanotechnology instantiates the science-fictionizing of the world.

Donna Haraway has argued that the science-fictionizing of technoculture, or the postmodern revelation that "the boundary between science fiction and social reality is an optical illusion," gives rise to a "cyborg" epistemology threatening humanistic borders.[30] Similarly, Scott Bukatman sees the new subjectivity created by the science fictions of technoculture as a "terminal identity," writing that "[t]erminal identity is a form of speech, as an essential cyborg formation, and a potentially subversive reconception of the subject that situates the human and the technological as coextensive, codependent, and mutually defining."[31] Haraway and Bukatman suggest that cyborg fusions and science fiction technologies transfigure embodied experience, enabling the appearance of a posthuman subject that N. Katherine Hayles describes as "an amalgam, a collection of heterogeneous components, a material-informational entity whose boundaries undergo continuous construction and reconstruction."[32] I argue that nanotechnology is an active site of such cyborg boundary confusions and posthuman productivity, for within the technoscapes and dreamscapes of nanotechnology, the biological and the technological interpenetrate, science and science fiction merge, and our lives are rewritten by the imaginative gaze—the new "nanological" way of seeing—resulting from the splice. The possible parameters of human subjectivities and human bodies, the limits of somatic existence, are transformed by the invisible machinations of nanotechnology—both the nanowriting of today and the nanoengineering of the future—facilitating the eclipse of man and the dawning of the posthuman condition.

Nanotechnology as Science, or, The Nanorhetoric

Nanotech is a vigorous scientific field anticipating a technological revolution of immense proportions in the near future, and Eric Drexler is right at the vanguard. Founder and chairman of the Foresight Institute as well as a research fellow at the Institute for Molecular Manufacturing, his scientific credentials (Ph.D. from MIT, a former visiting appointment at Stanford, numerous publications) are impressive. But Drexler's seminal and influential *Engines of Creation,* outlining his program for nanotech research, is composed as a series of science-fictional vignettes. From spaceships to smart fabrics, from *A.I.* to immortality, *Engines of Creation* is a veritable checklist of science-fictional clichés—Drexler's insistence on scientificity notwithstanding—and the narrative structure of the book unfolds like a space opera: watch as brilliant nanoscientists seize control of the atom and lead humankind across the universe . . . and beyond!

The operatic excess of nanowriting—that genre of scientific text in which the already inevitable nanotech revolution can be glimpsed—characterizes even technical publications by Drexler, Ralph Merkle, Markus Krummenacker, Richard Smalley, Daniel Colbert, Robert Freitas, Jr., J. Storrs Hall, and other prophets of the nanofuture. Speculative and theoretical, these texts demonstrate what is possible but not what has been accomplished, what has been successfully simulated but not what has been realized (for example, Merkle writes that nanoscientists are working diligently to "transform nano-technology from computer *models* into *reality*" [emphasis added][33]). These texts frame their scientific arguments with vivid tales of potential applications, which are firmly the stuff of the golden age of science fiction. Matter compilers, molecular surgeons, spaceships, space colonies, cryonics, smart util-ity fogs, extraterrestrial technological civilizations, and utopias abound in these papers, borrowing unabashedly from the repertoire of the twentieth century science-fictional imagination.[34]

Consequently, the experimental evidence supporting the reality of nanotech has been marshaled into battle to divide the science from its "sci-fi" associations. Nanotechnology is a realistic science, many researches claim, because biological "nanomachines" like enzymes and viruses already exist in nature; there is no reason, then, why human engineers could not construct similar molecular devices.[35] Unfortunately, even with nature as a model, the tangible products of nanoresearch are extremely preliminary. The more celebrated experimental results, in no particular order, include: 1) Engineered proteins and synthetic molecules with protein-like capabilities (William DeGrado and colleagues accomplished the former in 1988; Donald Cram, Jean-Marie Lehn, and Charles Pederson shared a Nobel Prize in 1987 for the latter); 2) An organic molecule pinned to a surface with a scanning tun-neling electron microscope (STM) (led by John Foster at IBM in 1988);[36] 3) The widely-publicized construction of the IBM logo on a silicon chip by pushing individual xenon atoms with an STM (led by Donald Eigler at IBM in 1989);[37] 4) The production of fullerines (earning Richard Smalley, Robert Curl, and Sir Harold Kroto a Nobel Prize in 1996) and their applications such as "nanopencils" that deposit molecular ink and increasingly precise manipulation of individual atoms;[38] and 5) Invented nano-novelties, such as a "nanoabacus" (produced in 1996 by an IBM team led by James Gimzewski), a "nanotrain" (a large mobile molecule crawling along a molecular "track," synthesized by Viola Vogel), and various rotating molecular motors.[39] These technical accomplishments, as laudable and fascinating as they are, do not represent the successful arrival of molecular manufacturing; nonethe-less, because they seem to suggest progression towards the "full potential" of nanotech, nanorhetori-cians maintain that the "evolution of nanotechnology" is a scientifically valid expectation.

Further evidence that nanotechnology is a real science, rather than a misguided fad, comes from its many signs of protodisciplinarity. The fact that professional scientists are actively working and staking their reputations on it is sociologically significant, and the visible confrontation between various nanotech research programs seeking to shape the field is symptomatic of the efforts of nanotechnology as a whole to attain the status of an acknowledged professional discipline.[40] These agonistic struggles within the technoscape have stabilized a field-specific lexicon as well as institu-tional structures—marked research funds, industrial conferences and university programs—supporting nanotech research. Drexler taught an engineering course on nanotechnology at Stanford in 1989, and such curricular inclusion supposedly indicates the belated recognition of an already exciting field, for Drexler writes, "At Stanford, when I taught the first university course on nanotechnology, the room and hallway were packed on the first day, and the last entering student climbed through a window."[41] A certain pedagogical credibility stems from Drexler's textbook on nanotech engineering and design called *Nanosystems: Molecular Machinery, Manufacturing, and Computation* (1992).[42] A textbook is usually at the trailing end of a scientific discipline rather than the forefront, but Drexler composed this tome, filled with the differential equations, quantum mechanical calculations and structural diagrams that had been missing from his earlier publications, seemingly with the intent of legitimating an increasingly maligned science. Since 1989, the Foresight Institute has sponsored annual international conferences on nanotechnology, bringing in researchers from all over the world. The first nanotech start-up company, Zyvex, appeared in Richardson, Texas in 1997 intending to develop nanodevices like Drexler's assembler in less than a decade.[43] Zyvex has been followed by a boom in nanotech interest in the Silicon Valley and other regions where industrial speculation and venture capital abundantly flow. There are even scholarly journals, such as *Nanotechnology* and *Nano Notes,* that publish exclusively the cutting-edge research in the field.

So it certainly looks like a science, and the people promoting the field are really trying hard to show why it is not science fiction. The main argument enforcing this division emerges, again, from the logic of the real versus the simulacrum; specifically, nanowritings insist that their visions of the future are grounded in "real science," while those futures described in science fiction are not. Take, for example, Drexler's comments on science fiction in *Engines of Creation:*

> By now, most readers will have noted that this [nanotechnology] . . . sounds like science fiction. Some may be pleased, some dismayed that future possibilities do in fact have this quality. Some, though, may feel that "sounding like science fiction" is somehow grounds for dismissal. This feeling is common and deserves scrutiny.
>
> Technology and science fiction have long shared a curious relationship. In imagining future technologies, SF writers have been guided partly by science, partly by human longings, and partly by the market demand for bizarre stories. Some of their imaginings later become real, because ideas that seem plausible and interesting in fiction sometimes prove possible and attractive in actuality. What is more, when scientists and engineers foresee a dramatic possibility, such as rocket-powered spaceflight, SF writers commonly grab the idea and popularize it.
>
> Later, when engineering advances bring these possibilities closer to realization, other writers examine the facts and describe the prospects. These descriptions, unless they are quite abstract, then sound like science fiction. Future possibilities will often resemble today's fiction, just as robots, spaceships, and computers resemble yesterday's fiction. How could it be otherwise? Dramatic new technologies sound like science fiction because science fiction authors, despite their frequent fantasies, aren't blind and have a professional interest in the area.
>
> Science fiction authors often fictionalize (that is, counterfeit) the scientific content of their stories to "explain" dramatic technical advances, lump them together with this bogus science, and ignore the lot. This is unfortunate. When engineers project future abilities, they test their ideas, evolving them to fit our best understanding of the laws of nature. The resulting concepts must be distinguished from ideas evolved to fit the demands of paperback fiction. Our lives will depend upon it.[44]

I have quoted this passage at length because of it several remarkable qualities intended to rescue nanotechnology from the ghetto of science fiction. While the first paragraph begins the radical task of reconciling science and science fiction, juxtaposing the languages of "possibility" and "fact," Drexler quickly departs from this goal and instead firmly separates science, and particularly nanotechnology, from the "fantasies" of fiction. He clarifies the assumed directional flow of reality into fiction: when science fiction is "real" the writer either landed on reality by chance or "grabbed" the idea from science. Drexler thus distinguishes science fiction writers from "other writers" and "engineers" who "examine the facts" (presumably Drexler fits into this category). He employs the idea of the "counterfeit" to describe science fiction—not, of course, citing Baudrillard, but drawing on the same understanding of the counterfeit as similar to but distinct from reality. He divides "our best understanding of the laws of nature" (Drexler's writing) from "the demands of paperback fiction" (science fiction), concluding that, because of the dangerously real consequences made possible by nanotech, our very lives depend on maintaining this division! What further rationale for recognizing the barrier between science and science fiction could one need?

Thus Drexler seemingly secures his work as science, but another tactic deployed by defenders of nanotech is to exclude Drexler and his sympathizers from the technoscape entirely. This strategy acknowledges and foregrounds the intractable science-fictionalisms of Drexler's science and thereby pronounces him a pariah, in effect preserving the rest of nanotech as "real science."[45] For example, Donald Eigler (of the xenon IBM logo) has audaciously declared that "[Drexler] has had no influence on what goes on in nanoscience. Based on what little I've seen, Drexler's ideas are nanofanciful notions that are not very meaningful."[46] Mark Reed, nanoelectronics researcher and Professor of Engineering and Applied Science at Yale, has said, "There has been no experimental verification for any of Drexler's ideas. We're now starting to do the real measurements and demonstrations at that scale to get a realistic view of what can be fabricated and how things work. It's time for the real nanotech to stand up" (emphasis added).[47] The force of this argument comes from the deluge of the "real," which, repeated *ad nauseam*, appears to drown Drexler and friends and engulf them in the irrationalities of their nanodreams. Again we see the rhetorical establishment of a powerful dichotomy of science versus science fiction, but this time constructed within the technoscape itself.

A final tactic used by nanorhetoricians, both Drexlerians and Drexler-detractors, is the oft-repeated story about the genesis of nanotech. I will call this foundational narrative the "Feynman origin myth." The story goes (and it is told by nearly everyone researching in this field, posted on their Web pages and repeated in their publications) that on December 29, 1959, Richard Feynman delivered a talk entitled "There's Plenty of Room at the Bottom" to the American Physical Society at the California Institute of Technology. Here, Feynman suggested the possibility of engineering on the molecular level, arguing that the "principles of physics, as far as I can see, do not speak against the possibility of maneuvering things atom by atom. It is not an attempt to violate any laws; it is something, in principle, that can be done."[48] Feynman further asserted that something like nanotech is "a development which I think cannot be avoided." Quotations and paraphrases of these statements run rampant throughout the discourse network as arsenal in the war to legitimate nanotechnology.[49] Such recourse to Feynman's speech has given rise to the belief that Feynman originated, authorized, and established nanotechnology. Assertions like "This possibility [of nanotechnology] was first advanced by Richard

Feynman in 1959"[50] and "Richard Feynman originated the idea of nanotechnology, or molecular machines, in the early 1960s"[51] are commonplace and have taken on the status of truisms. Feynman's talk is continually invoked to prove that nanotechnology is a real science, but not because of the talk's theoretical, mathematical, or experimental sophistication; indeed, judging from the language used—the numerous appearance of "possibility," "in principle," "I think," and the telling "it would be, in principle, possible (I think)"—it is clear that Feynman's talk was just as speculative as (if not more than) any article penned by Drexler, Merkle, or their associates.

The Feynman origin myth is resurrected over and over again as a cheap way of garnering scientific authority. How better to assure that your science is valid than to have one of the most famous physicists of all time pronouncing on the "possibility" of your field? It is not uncommon for nanorhetoricians, when referencing the talk, to remind their audience that Feynman won the 1965 Nobel Prize in physics. Merkle candidly reveals that name recognition and cultural capital are the main values of this tactic when he writes: "One of the arguments in favor of nanotechnology is that Richard Feynman, in a remarkable talk given in 1959, said that, 'The principles of physics, as far as I can see, do not speak against the possibility of maneuvering things atom by atom.'"[52] The argument is clearly not *what* Feynman said, but "is that" *he* said it. The argument hinges on Feynman's unique vision, what he "can see," something special about Feynman's scientific ability that transforms a speculative statement into a description of reality. A frank example of fetishizing the author and the origin (the Foresight Institute even offers a "Feynman Prize"), Feynman's talk grounds nanotechnology not in the real but in authoritative discourse. Nevertheless, the Feynman origin myth is perceived as dissociating nanotechnology from science fiction.

To its credit, nanotech has been fairly successful in the battle to vindicate itself as a real science, as something very different from science fiction despite how much it may seem like science fiction. The anti-SF rhetoric has even made its nanodreams appear more like inevitabilities to a larger audience. From 1992, when Drexler and company unveiled a wonderful nanofuture to the U.S. government and achieved the allocation of special NSF funds for nanoscale research, to the implementation of the 2001 National Nanotechnology Initiative, the foundations for which grew out of Congressional testimonies by Smalley, Merkle, and other key figures in the field, nanorhetoric triumphed in transforming science fiction visions into manifest and lucrative national ventures.[53] Even President Clinton, announcing the National Nanotechnology Initiative at Caltech on January 21, 2000, demonstrated his absorption of nanorhetoric by citing the 1959 Feynman talk, along with several imaginary coming attractions of the nanofuture, as evidence for the decisive role that nanotechnology will play in bringing about an "era of unparalleled promise."[54] Thus despite many determined critics, nanotech managed to secure its professional future by combining fantastic speculation with concerted attacks on science fiction. Indeed, considering nanotech's rapid expansion in academia and industry, the reputable scientists involved, and its current high profile, there appears little doubt that nanotech is real science.

However, the "sci-fi" anxieties haunting defenders of nanotechnology disclose its scandalous proximity to science fiction, and, I argue, only rhetoric is maintaining the separation. Furthermore, I will show that this rhetoric thoroughly deconstructs itself in a futile struggle for boundary articulation that has already been lost.

Nanotechnology as Science Fiction, or, Deconstructing the Nanorhetoric

Recall Drexler's arguments regarding science fiction. Drexler must explicitly distinguish his science from paperback fiction because his nanonarratives borrow extensively from pre-existing genre conventions. Drexler's stories—like those found throughout nanowriting—describe the world transformed by imagined feats of science and engineering relegated to the unspecified future, and even when denying the science-fictionality of his vignettes by emphasizing that they are "scientifically sound," Drexler cannot avoid drawing attention to the fact that they do, after all, "sound like science fiction." Although Drexler confirms the conventional assumption that science is the real, science fiction its imaginary simulacrum, when he says that his science "sounds *like*" fiction, he reverses the assumed order. Science fiction has anticipated science, and the ensuing science is not ultimately delineated from science fiction by Drexler's arguments.

Though Drexler distinguishes science fiction writing from his kind of writing through the criterion of mimesis, science fiction writers who "grab the idea [from science] and popularize it" are not logically different from writers who "examine the facts" of science and popularize them, as *Engines of Creation* is intended to do. Along the same lines, the criterion that Drexler's stories are scientifically sound while science fiction stories are (presumably) not is challenged when he acknowledges that science fiction "imaginings" frequently "become real" (again reversing the assumed order). Science and science fiction dynamically and frequently shift structural positions in Drexler's writing, both suggested to be inhabited by "the real" at the same time as each paradoxically appears to simulate the other. That is to say, the real has become simulation and the simulation has become real.

None of these inconsistencies mean that Drexler is not writing good science; they do mean that the boundary between science fiction writers and writers of what Drexler calls "theoretical applied science," like himself, is hopelessly blurred. Tellingly, Drexler has personally forayed into the production of genre science fiction texts, writing an introduction to the short story collection, *Nanodreams* (1995), where he discusses the importance of science fiction in assessing future technologies.[55] The unavoidable failure of the dichotomy between science and science fiction occurs when Drexler, having apparently given up the endeavor, calls the scenarios described in *Engines of Creation* "science fiction dreams."[56]

Thus the division between writers of science fiction and writers of "theoretical applied science" or "exploratory engineering" is destabilized and confused. "Scientifically sound," according to Drexler, can be a quality of both kinds of writing—destroying the criterion, erasing the division. Ultimately, Drexler's nanowriting indicates that science fiction precedes and supersedes "its" science, echoing Baudrillard's "precession of simulacra": the simulacra coming before, displacing and supplanting, making the real seem to be the not-real, the science to be the science-fictional.[57]

Determining that Drexler's version of nanotechnology is inseparable from its science-fictionalisms would apparently make the nanotactic of excluding him from the field more effective. After all, if his writing is indeed science-fictional, then he is not, according to Reed, part of "the real nanotech." However, attempts to banish Drexler from the field he established actually have the ironic effect of highlighting the science-fictionality of nanotech. When Eigler states that Drexler "has had no influence on what goes on in nanoscience," he is disregarding Drexler's seminal technical publications and the considerable contributions of his Foresight Institute; furthermore, Eigler is in flat contradiction to the vast expanses of the technoscape recognizing Drexler's inspiring influence[58]—including Smalley, who says that Drexler "has had tremendous effect on the field through his books."[59] When Reed says that Drexler's ideas have not been experimentally verified and therefore are not part of the "real" nanotech, he is disregarding the validity of all theoretical science—clearly a problematic move. Consequently, Drexler cannot be so simply exiled: he has persuaded not only individual nanoscientists but also governmental funding boards about the inevitable nanofuture,[60] and accordingly, nanotechnology should acknowledge the heavy speculation that remains fundamental for its own development as a research field. After all, having proclaimed that Drexler is "science fictional" and "not real," yet ultimately obliged to recognize his influence, this tactic to expel science fiction from science backfires on itself.

Even Merkle's *response* to these exclusionary efforts eventually backfires. In a letter to the editor of *Technology Review,* he writes:

> While I am happy to see the increasing interest in nanotechnology, I was disappointed by your special report on this important subject. Mark Reed summarized one common thread of the articles when he said "There has been no experimental verification for any of (Eric) Drexler's ideas." Presumably this includes the proposal to use self-replication to reduce manufacturing costs. The fact that the planet is covered by self-replicating systems is at odds with Reed's claim.

> Self-replicating programmable molecular manufacturing systems, a.k.a. assemblers, are not living systems. This difference lets Reed argue that they have never before been built and their feasibility has not been experimentally verified. Of course, this statement applies to anything we have not built. Reed has discovered the universal criticism. Proposals for a lunar landing in 1960? Heavier-than-air flight before the Wright brothers? Babbage's proposal to build a computer before 1850? No experimental verification. Case closed.[61]

Merkle musters a "fact" (i.e. that self-replicating systems abound in nature) in support of Drexler and builds an argument for the validity of scientific speculation, successfully countering Reed's implication that Drexler's science is not "real." Drexler is salvaged, put back on the secure ground of reality. But while accomplishing Drexler's reassimilation into the field, Merkle also winds up equating nanotechnology with science fiction. Merkle suggests that nanotechnology is a real science, even though it lacks experimental verification, because proposals for a lunar landing in 1960, considerations of heavier-than-air flight before the Wright brothers, and Babbage's idea for a computer had no experimental verification and yet these ideas eventually found verification after time. "Case closed," he writes. But, of course, speculations for a moon voyage, for heavier-than-air flight, and for computers of various sorts had existed long before their "real" incarnations—think of the stories of Jules Verne, H. G. Wells, Hugo Gernsbeck, Isaac Asimov, Robert Heinlein, Arthur C. Clarke and countless others—all of which were and still are clearly marked as science fiction. Thus in recuperating the speculations of nanowriting, Merkle solidifies the relay between nanotechnology and science fiction. Before moon voyages, air flight, and computers there was science fiction; before the nanotechnology revolution of the future there is the anticipatory nanotechnology of today. Nanotechnology is science fiction. Case closed?

This dissolving boundary between science and science fiction in nanowriting elsewhere occurs as intertextuality, in the sense that loci of meaning within nanowritings are frequently dependent upon a larger web of texts, both science and science fiction, that enable their signification. In this respect, nanowritings are what Jonathan Culler describes as "intertextual constructs" that "can be read only in relation to other texts, and [they are] made possible by the codes which animate the discursive spaces of a culture."[62] For example, the concept of the "Diamond Age"—describing how the nanotechnology era will be historicized relative to the Stone Age, the Bronze Age, the Silicon Age, etc.—appears in science fiction, particularly Neal Stephenson's nanotech novel, *The Diamond Age* (1995), and in Merkle's *Technology Review* survey article, "It's a Small, Small, Small, Small World" (1997).[63] Each text, science and science fiction, assumes reader familiarity with the terminology deployed by the other.

Stephenson's novel, similarly, describes a "Merkle Hall" located within the nanotech corporation, Design Works, whose ceiling, reminiscent of Michelangelo's Sistine Chapel, is covered with a fresco depicting the pantheon of nanotech, wherein Feynman, Merkle, and Drexler mingle with more fictional personalities.[64] Fact and fiction merge in the blender of nanowriting where allusions are creatively drawn from both technical reports and popular novels.

The issue of science-fictional allusion arises even more strikingly in J. Storrs Hall's theoretical elaboration of a nanotech "utility fog"—an engineered, pervasive substance for complete environmental control and universal human-machine interface.[65] Hall's essay in exploratory engineering, diffusely impregnated with science fiction tropes, is structured around witty references to many canonical science fiction texts, including *Forbidden Planet* (1956), Robert Heinlein's "The Roads Must Roll" (1940), Jules Verne's *From the Earth to the Moon* (1865), H. G. Wells's *The Shape of Things to Come* (1933), and Karl Capek's *R.U.R.* (1920), suggesting that nanotechnological thinking is essentially a process of writing from the margins of other fictional futures, other textual worlds. Within nanowriting, the facile permeability of these worlds of science and fiction, the ease with which concepts and signs traffic between them, challenges any stringent boundrification. The tactics of separating nanotech from the science fiction with which it is complicit fail on every level.

As a final bit of evidence, let's return to the Feynman origin myth. Despite nanorhetoricians' frequent citations of the talk to support the realness of their discipline, the talk itself sits awkwardly with such a purpose. We have seen the indeterminacy and speculative nature of the language Feynman uses, and strikingly, the talk is composed as a series of science fiction stories, just like Drexler's *Engines of Creation*. Feynman tells stories about tiny writing, tiny computers, the actual visualization of an atom, human surgery accomplished by "swallow[ing] the surgeon," and "completely automatic factories"—certainly not impossibilities, but still the conceits of numerous genre science fiction narratives long before Feynman stepped to the podium. Thoroughly penetrated by the science-fictional imaginary, it is no coincidence that Feynman's nanotech looks just like Drexler's nanotech, fabricated from the same "science fiction dreams."

The Feynman origin myth thus contains in itself the deconstruction of the nanotech/science fiction dichotomy. The cavalier way in which the myth is used by both Drexlerians and those who challenge Drexler's vision of nanotech is further indication of its deconstructive tendencies. Consider, for example, the response of Thomas N. Thetis (IBM Research Division) to the *Technology Review* article where Reed implies that Drexler's nanotech is not real: "Congratulations on your review . . . Your writers clearly distinguished hype from hard science and vision from reality. I was reminded of

Richard Feynman's famous 1959 after-dinner talk . . . Feynman managed to foreshadow decades of advances . . . I know that his vision influenced at least a few of the individuals who have made these [hard science] things happen."[66] That Thetis can speak of "vision" opposed to "reality" in one sentence and of Feynman's "vision" that *contributed* to hard (i.e. real) science in another reveals the ease of appropriating such a myth for one's own purposes, the impossibility of simply excluding Drexler's "vision" from the field, and the blurring of science and science fiction within the Feynman talk. After all, if vision is opposed to reality, then Feynman's talk abandoned reality entirely.

Even as a genesis story, the Feynman myth only succeeds in making a science fiction of nanotechnology. Nanotechnology is supposedly a real science because it was founded and authorized by the great Richard Feynman. But this origin is not an origin, and its displacement unravels the structure of its legacy. The Feynman myth would only work if it clearly had no precedents, if it was truly an "original" event in intellectual history, if even Feynman had offered a unique, programmatic conception of how nanotechnology was to be accomplished. Yet this is not the case: Feynman merely depicted a speculative vision of a possible technology, and science fiction writers, as they have done with so many things, had already beaten him there. Technologies or theoretical concepts that are identifiably similar to current visions of nanotechnology appear in Theodore Sturgeon's "Microcosmic God" (1941), Robert Heinlein's "Waldo" (1942), Eric Frank Russell's "Hobbyist" (1947), James Blish's "Surface Tension" (1952), and Philip K. Dick's "Autofac" (1955), all well before Feynman gave his now-mythical talk.

Although there is no evidence that Feynman personally read any of these science fiction stories, his friend Albert R. Hibbs (senior staff scientist at the Jet Propulsion Laboratory) did read "Waldo" and described it to Feynman in the period just before Feynman composed his talk.[67] And indeed, Heinlein's influence haunts Feynman's depiction of nanotechnology. In Heinlein's novella, the eponymous genius, Waldo, has invented devices–known as "waldoes"–which are mechanical hands of varying sizes, slaved to a set of master hands attached to a human operator. Heinlein writes that the "secondary waldoes, whose actions could be controlled by Waldo himself by means of his primaries," are used to make smaller and smaller copies of themselves ("[Waldo] used the tiny waldoes to create tinier ones"), ultimately permitting Waldo to directly manipulate microscopic materials by means of his own human hands.[68] Heinlein thus hypothesizes a method for molecular engineering that Feynman in his talk, without crediting his source, offers as a means to "arrange the atoms one by one the way we want them." Feynman describes his proposed system:

> [It would be based on] a set of master and slave hands, so that by operating a set of levers here, you control the "hands" there . . . I want to build . . . a master-slave system which operates electrically. But I want the slaves to be made especially carefully by modern large-scale machinists so that they are one-fourth the scale of the "hands" that you ordinarily maneuver. So you have a scheme by which you can do things at one-quarter scale anyway–the little servo motors with little hands play with little nuts and bolts; they drill little holes; they are four times smaller. Aha! So I manufacture [with these hands] . . . still another set of hands again relatively one-quarter size! . . . Thus I can now manipulate the one-sixteenth size hands. Well, you get the principle from there on.[69]

The originality of the Feynman myth crumbles, for we can see that Feynman's talk emerges from genre science fiction. Feynman's method of molecular manipulation is borrowed from Heinlein. Even the proposition for internal microscopic surgery—a notion Feynman credits to Al Hibbs—was already proclaimed as an "original" idea by Heinlein in the "Waldo" novella, where Heinlein writes that microscopic surgery via microscopic machines "had never been seen before, but Waldo gave that aspect little thought; no one had told him that such surgery was unheard-of."[70] The mythologized order of precedence is therefore reversed, for it becomes evident that speculations of nanotech were freely circulating in the discourse of science fiction long before science "grabbed the idea." If we really want to locate an origin to nanotechnology, it is not to Feynman that we must look, but to science fiction.

Consequently, I reiterate that in the case of nanotech we have a situation where simulation has preceded and enveloped "real" science, where the line between science and science fiction is blurred, made porous, and effaced. It even seems likely that this hybridity has been responsible for nanotech's recent financial success; companies have been founded and government officials have been awed less by nanotech's real accomplishments—for there are few—but rather on its dream of the future, its promise of a world reborn: its science fiction indistinguishable from its science. Rapidly becoming a major actor in the science-fictionizing of technoculture—along with certain other interstitial sciences and technologies, such as virtual reality, cybernetics, cloning, exobiology, artificial intelligence and artificial life—nanotechnology exerts strong symbolic influence over the way we conceptualize the world and ourselves. In other words, as a science (fiction) with enormous cultural resources and increasing historical significance, nanotechnology claims for itself a powerful role in the human future and the future of the human.

Posthuman Engineering

The birth of nanotechnology as a scientific discipline provokes the hyperreal collapse of humanistic discourse, puncturing the fragile membrane between real and simulation, science and science fiction, organism and machine, and heralding metamorphic futures and cyborganic discontinuities. In both its speculative-theoretical and applied-engineering modes, nanotechnology unbuilds those constructions of human thought, as well as those forms of human embodiment, based on the security of presence and stability—terrorizing presentist humanism from the vantage point of an already inevitable future. As Jacques Derrida has repeatedly suggested, the deconstruction of metaphysical structures allows us to "pass beyond man and humanism, the name of man being the name of that being who, throughout the history of metaphysics or of ontotheology—in other words, throughout his entire history—has dreamed of full presence, the reassuring foundation, the origin and the end of play."[71] Critiquing humanism from within while simultaneously stepping radically outside of the domain defined by humanism opens a subject position other than that implanted between essence and eschatology—which is the position of the human, for the "name of man has always been inscribed in metaphysics between these two ends."[72] With a similar agenda, Michel Foucault has argued for the historic boundaries of humanism, depicting an epistemic closure marking the end of man as an entity: "As the archaeology of our thought easily shows, man is an invention of recent date. And one perhaps nearing its end."[73] The intellectual breakdown of humanism is advanced through the collision between human flesh and postmodern technologies, where the relational interface mediates the emergence of new posthuman haptic spaces—machinic, virtual, material, and meaty—as Paul Virilio, Brian Massumi, and N. Katherine Hayles have suggested.[74] I argue that nanotechnology participates in the techno-de(con)struction of humanism, forcing us to think otherwise through its ambiguous hyperreal status and its narratives of corporeal reconfiguration from beyond the temporal horizon, fabricating new fields of embodiment and facilitating our becoming posthuman by envisioning a future where the world and the body have been made into the stuff of science fiction dreams.

Kelly Hurley has written that posthuman narratives of "bodily ambiguation" and "speculations on alternate logics of identity that rupture and exceed the ones we know" restructure our somatic experiences, for these posthuman narratives work to "disallow human specificity on every level, to evacuate the 'human subject' in terms of bodily, species, sexual, and psychological identity," supporting the "generation of posthuman embodiments both horrific and sublime."[75] Nanotechnology produces such narratives of bodily ambiguation and articulates an alternative logic of identity—a subversive technoscientific gaze that I will term "nanologic"—in the stories of nanofutures circulating within the technoscape and beyond. (Indeed, nanoscientists seem to align with Hurley on the immediate tangible impact of posthuman narratives in their suspicion that the world has already been remade by nanotech, that nanowriting's extrapolation of possible posthuman futures necessitates the "foresight" that Drexler and others have been advocating since their earliest publications.)[76]

Whether utopian visions or catastrophic nightmares, nanonarratives resist traditional humanist interpretations by repeatedly depicting the future in terms that disequilibrate the human body. From the eroticized collective consciousness of the Drummers in Stephenson's *The Diamond Age* (1995), to the lycanthropic transformations of Dean Koontz's *Midnight* (1989), to the permeability of "enlivened" city-structures and body-structures in Kathleen Ann Goonan's *Queen City Jazz* (1994), to the metamorphosis of the entire human population into billowing sheets of sentient brown sludge in Greg Bear's *Blood Music* (1985), posthuman bodies in nanonarratives are never stable, never idealized, never normative, never confined; the limits of posthuman corporeality are as wide as the nanological imagination. Nanologic disrupts the boundaries and the configurations of the human body, rebuilding the body without commitment to the forms given by nature or culture; and thus nanotechnology, as both a contemporary discourse and a future material science, is an instrument of posthuman engineering.

Rather than purveying a posthumanism in which the subject is in danger of losing the body—an imagined fate that Hayles has extensively critiqued[77]—nanonarratives articulate posthuman subjectivities resulting from embodied transformations. Embodiment is fundamental to nanonarratives because, in the science of nanotechnology, *matter* profoundly *matters*. Nanotech respects no unitary construct above the atom and reduces everything to pure materiality, demolishing metaphysical categories of identity. Accordingly, nanologic does not support any sort of abstracted, theoretical construction of the body because nanotech unbounds the body, puts its surfaces and interiors into constant flux. The posthuman bodies conditioned by nanologic are therefore always individuated experiences of embodiment in an endless array of possible bodily conformations, where all borders are fair game.

Nanologic is a cyborg logic, imploding the separation between the biological and the technological, the body and the machine. As we have seen, one of the arguments legitimating nanotechnology is that biological machines like ribosomes and enzymes and cells are real, and consequently there is nothing impossible about engineering such nanomachines. But the very ease in describing biological objects as machines indicates the cyborgism of nanotech, its logic of prosthesis, its construction of bodies and machines as mutually constitutive. Nanotechnology envisions the components of the body and mechanical objects as indistinguishable, and, subsequently, utilizes the biological machine *as the model* for the nanomachine, achieving a terminal circularity. Nanologic removes all intellectual boundaries between organism and technology—as Drexler puts it, nanologic causes "the distinction between hardware and life . . . to blur"[78]—and human bodies become posthuman cyborgs, inextricably entwined, interpenetrant, and merged with the mechanical nanodevices *already inside of them*.

Having become cyborganic machines, bodies in the grasp of nanologic can be reassembled or reproduced with engineering specificity. Unlike genomic cloning, which merely provides genotypic but not necessarily phenotypic identity, the copying fidelity of nanotechnology is so exact that copies would have precise identity down to the atomic level. Feynman (following Heinlein) foresaw this in his talk: "all of our devices can be mass produced so that they are absolutely perfect copies of one another."[79] The ability of nanodevices to produce exact copies—copies of themselves, copies of their constructions—is fundamental to nanologic, and it is not, perhaps, entirely a coincidence that for more than a decade Merkle directed the groundbreaking Computational Nanotechnology Project for Xerox.[80] The potential for nanotechnology to reproduce anything exactly, accurate in every atomic detail, or to reconstruct anything into an identical copy of anything else, leads to posthuman nanonarratives that, undermining our conceptions of identity and origin(ality), need not become literalized to have transformed the architectures of our somatic experience. As Hurley suggests, posthuman narratives ask us "to imagine otherwise, outside the parameters of 'the human'," thereby opening up new possibilities of corporeality that change the way we conceive ourselves.[81] Such possibilities are illustrated by the following series of nanoscenarios:

○ A wooden chair, subjected to a herd of nanobots, can be transformed into a table, its "chairness" subtly and efficiently morphed into "tableness." Nanologic undermines essentialism, insisting that every "thing" is simply a temporary arrangement of atoms that can be endlessly restructured.

○ A wooden chair can be transformed into a living fish. There is no magic here, merely a precise rearrangement of molecules. Life instantly arises from dead material; as Drexler writes, nanologic reveals that "nature draws no line between living and nonliving."[82]

○ A wooden chair can be transformed into a human (i.e. Homo sapiens). The same process for the fish now challenges humanist metaphysics a little more forcefully. The resulting human could even be a specific person like Sigourney Weaver (posthuman icon from the *Alien* films), identical to the movie star in every respect: DNA, proteins, phospholipids, neurotransmitters, memories.

○ A fish can be transformed into a human. The fish does not die, does not stop being, it merely becomes human.

○ A human, subjected to a herd of nanobots carrying the data set for another human, can suddenly become someone else. Human *A* and Human *B* share the same matter, the same coordinates in space-time; although they have different identities, although they are different people, they are the same being.

○ A woman can be metamorphosed into a man, or vice versa, or in various partial combinations. Mono-, inter- and transsexuality can be manifested in a single figure. Tissues, hormones and chromosomes can be refabricated. The posthuman body is thus queered: sex and sexuality made infinitely malleable, sexual difference slipping into sexual indeterminacy, or deferral.

○ A human body can become the copy of an already existing human body. Say, for example, Harrison Ford (posthuman icon from *Blade Runner)* transforms into Sigourney Weaver. Then there are two Sigourneys, identical down to the memories, even down to the belief that each is Sigourney Weaver and the other is the copy. There is no possible way of telling them apart, no possible way of telling which was the "original." Someone might ask, "Will the real Sigourney please stand up," but inevitably they both will. More disturbing than clones or even the android replicants in *Blade Runner,* which merely mimic, these nanocopies actually *are.* Nanologic again destroys the difference between real and simulacrum.

○ Nanotechnology can devise a matter-transporter to facilitate human travel across great distances of space.[83] At one end, nanobots dismantle the human traveler atom by atom, recording the location of each molecule, until the traveler is just a pile of disorganized material. The nanomachines feed data into a computer system, which instructs another group of nanobots at the terminal end of the transporter, working from a feed of appropriate elements, to reassemble the human traveler exactly as he or she had been at the proximal end. The traveler will have no memory of the trip but will emerge precisely as he or she was when the process began; though made from different atoms, the traveler is still the same person. Embodiment has been distributed across a spatial divide and between separate accumulations of matter. Furthermore, the data can be reused to construct multiple, identical copies of the traveler. Personhood can be duplicated, flesh Xeroxed, minds mimeographed.

○ Human bodies can be modified well beyond the confines of experience, becoming alien formations or improbable mélanges. Nanotechnology empowers posthuman imaginations to achieve outlandish physical alterations. (How many tentacles would you like to have?)

○ Finally, nanologic enables us to think beyond human boundaries in a tragic sense, for nanotechnology can also bring about a post-human future where all of humanity has ceased to exist and nothing new emerges from the wreckage. This fate is made possible by insidious nanoweapons of mass destruction, or the nanocalyptic hypothesis of out-of-control nanobots turning the entire biosphere into "grey goo."[84] While providing a means to engineer new posthuman embodiment, nanotech also provides a means to engineer posthuman extinction.

As these scenarios suggest, nanotechnology has unprecedented effects on the way we are able to conceptualize our bodies, our biologies, our subjectivities, our technologies, and the world we share with other organisms. Whether positing the liberation of human potential or the total annihilation of organic life on this planet, nanologic demands that we think outside of the realms of the human and humanism. Nanologic makes our bodies cyborg and redefines our material experiences, redraws our conceptual borders, and reimagines our future. Accordingly, even before the full potential of a working nanotechnology has been realized, we have already become posthuman. Indeed, posthuman subjects abound in the nanoliterature, and although science fiction novels like Ian McDonald's *Necroville* (1994), James L. Halperin's *The First Immortal* (1998) or Michael Flynn's *The Nanotech Chronicles* (1991) imagine posthuman nano-modified bodies as appearing at some ambiguous point in the future, other "nonfictional" posthuman beings exist already, right now, within the popular and professional writings of nanoscientists. As real, embodied, material entities, enmeshed in the semiotics of nanologic, these posthumans are found at nanotechnology's intersection with cryonics.

Drexler, Merkle and other nanoscientists are deeply involved in the idea of freezing and preserving human bodies, or parts of human bodies, until the proper nanotechnology has been developed in the future that can revive and heal them. Freeze the body now and eventually nanotechnology will resurrect the subject, reversing not only the cellular damage caused by the freezing process, but also the damage that had originally caused the person to die, maybe even building an entirely new body for the cryonaut. Cryonic science is not simply tangentially related to nanotechnology, but has become a principle extension of nanologic—evidenced by the ubiquitous discussions of cryonics at all levels of nanodiscourse, from fanzines to university conferences.[85] Furthermore, Merkle is a director of the Alcor Life Extension Foundation, a cryonics institute founded in 1972, and he also hosts a cryonics web page; Drexler is on the scientific advisory board of the Alcor Foundation and has written extensively about cryonics in his books and scientific journal articles.[86]

Even in Drexler's first nanotech publication, cryonic resuscitation is evoked when Drexler writes that the "eventual development of the ability [of nanotechnology] to repair freezing damage [to cells] (and to circumvent cold damage during thawing) has consequences for the preservation of biological materials today, provided a sufficiently long-range perspective is taken."[87] Drexler thus implies that projected technologies of the future determine how we should deal with human tissues and human bodies in the present. Again nanowriting uses the language of the "already inevitable" and assumes that the full potential of nanotech has essentially been realized, temporal distance notwithstanding. Consequently, as deployed within the discourse of nanotechnology, the fact that cryonic techniques are currently in use means that nano-modified bodies are among us even now. Those who are dead but cryonically frozen have been encoded by nanologic as already revived, as already outside the humanistic dichotomy of dead/alive, as already voyagers into a brave new world of nanotech splendor . . . as already posthuman.

This nanological encoding of the cryonaut is evident when Drexler writes of cryonic resurrection in the science-fictional present tense, collapsing present and future, medical reality and technological fantasy, human death and posthuman revivification, into a single syntagmatic episode of *Engines of Creation*. Drexler tells of a hypothetical contemporary patient who

> has expired because of a heart attack. . . . [T]he patient is soon placed in biostasis to prevent
> irreversible dissolution. . . . Years pass. . . . [During this time, physicians learn to] use cell

repair technology to resuscitate patients in biostasis. . . . Cell repair machines are pumped through the blood vessels [of the patient] and enter the cells. Repairs commence. . . . At last, the sleeper wakes refreshed to the light of a new day—and to the sight of old friends.[88]

By way of alluding to H. G. Wells's *When the Sleeper Wakes* (1899), a canonic science-fictional depiction of sleeping into the future, Drexler validates and necessitates present-day acts of cryonic freezing within his prophecy of the coming nanoera. While indicative of nanowriting's dependence on the conventions of genre science fiction, this passage more significantly indicates how nanowriting's implosion of science into science fiction transmutes formerly human subjects into posthuman entities, amalgams of discourse and corporeality, biology and technology. For Drexler's cryonaut becomes posthuman at the moment of being incorporated into nanonarrative, thereby surviving its human death and becoming reborn through its cyborg interpenetration with nanomachines. And though the cryonaut in Drexler's story is hypothetical, other more specific cryonauts are made posthuman through the same mangle of nanologic.

Take, for example, Walt Disney—perhaps the world's most famous cryonically preserved character. In a wonderful semiotic tangle, the discourses of nanotechnology, cryonics, hyperreality and posthumanism all converge under the sign of Disney. Baudrillard has frequently written on the viral expansion of Disneyism, the "disnifying" of postmodern culture, the hyperreality of which Walt's own cryonic suspension is a telling symptom.[89] Bukatman expands on Baudrillard's depiction of the pervasive hyperreality of Disneyism, arguing that the "hypercinematic" architectures of Disneyesque spaces literally incorporate human bodies into their cybernetic systems, begetting cyborg terminal identities.[90] The Disney posthuman factory described by Baudrillard and Bukatman will be dramatically improved by the advent of nanotechnology, for nanoscientist and aerospace engineer Tom McKendree suggests that the "simulations" at Disneyland and other heightened realities will become even more of "a total experience" through nanotech's ability to "make the fantasies real."[91] Disneyism, already complicit with the reproduction of hyperreality and posthumanity, is simply attenuated by the imagineering capabilities of nanotechnology—so it is no mere coincidence that Disney "the man" becomes manifested at the point where nanologic merges with cryonics.

Consider Merkle's "It's a Small, Small, Small, Small World" essay: the title evokes the small world of atoms and assemblers purveyed by nanotechnology and, simultaneously, the "It's a Small World" ride at Disneyland and Disneyworld whose infectious and repetitious song ("It's a small world, after all! It's a small, small world!") metonymically stands for the Disneyscape as a whole. Disneyism is thus imported into nanowriting as metaphor for the nanoworld itself, and appropriately so—for not only does this figural resonance reveal the embeddedness of nanologic in the plane of hyperreality, where science and science fiction are one and the same, but furthermore, Walt's crystallized body is thereby absorbed into the Tomorrowland-like nanofuture that enables its return from the dead. Merkle details the coming "Diamond Age" of nanotechnology where the "ability to build molecule by molecule could also give us surgical instruments of such precision that they could operate on the cells and even the molecules from which we are made,"[92] and as many nanowriters have explained, such surgical precision will surely bring about cryonic resurrection.[93] Although Disney is presently on ice, waiting to be reborn through the advances of nanotechnology, within nanowriting—where a "small world" of quotidian miracles is deemed already accomplished, where "nanotechnology will inevitably appear regardless of what we do or don't do"[94]—Disney the sleeper already wakes. The future is now, and through the textual machinations of nanowriting that enable preserved human bodies to surmount their own deaths, Walt Disney himself has been transmuted into a posthuman creature of flesh, machines and hypersigns.

If nanologic's symbolic reprocessing of cryonauts like Walt Disney is any indication, then the transformation of the world envisioned by nanowriting is highly performative, and posthuman evolution has already begun. Accordingly, if nanotech is turning us posthuman, a critical scrutiny of the direction nanotechnology takes and an engaged involvement in the corresponding changes to our lives and our bodies is required to ensure that becoming posthuman is accomplished in our own terms. In *The Diamond Age,* Stephenson issues a note of caution as his novel replicates the narrative of nanotech inevitability, writing that "nanotechnology had made nearly anything possible, and so the cultural role in deciding what should be done with it had become far more important than imagining what could be done with it."[95] Nanotechnology empowers us to write our own posthuman future, but considering the massive biological, ecological, corporeal and cultural changes heralded by nanologic (be they utopic or apocalyptic), as voyagers into the future we must exercise the necessary foresight.

Indeed, foresight is a note that echoes throughout the technoscapes and dreamscapes of nanotechnology, from popular novels to experimental reports, as both a warning and an enticement. Haraway has similarly called for active intervention into the cyborg metamorphoses of our posthuman futures, writing that as "[a]nthropologists of possible selves, we are technicians of realizable futures."[96] Nanotechnology and all of its implications are on the horizon, bodied forth by the speculations of science and of fiction. With the nanofuture in sight, we must prepare for our posthuman condition . . . for it may be a small world, after all.

◯◯◯

Less is More: Much Less is Much More: The Insistent Allure of Nanotechnology Narratives in Science Fiction Literature

BROOKS LANDON

"Then you mean to say there is no such thing as the smallest particle of matter?" asked the Doctor.

"You can put it that way if you like," the Chemist replied. "In other words, what I believe is that things can be infinitely small just as well as they can be infinitely large. Astronomers tell us of the immensity of space. I have tried to imagine space as finite. It is impossible. How can you conceive the edge of space? Something must be beyond— something or nothing, and even that would be more space, wouldn't it?"

"Gosh," said the Very Young Man, and lighted another cigarette.

The Chemist resumed, smiling a little. "Now, if it seems probable that there is no limit to the immensity of space, why should we make its smallness finite? How can you say that the atom cannot be divided? As a matter of fact, it already has been. The most powerful microscope will show you realms of smallness to which you can penetrate no other way. Multiply that power a thousand times, or ten thousand times, and who shall say what you will see?"

⊚ Ray Cummings, "The Girl in the Golden Atom" (1919)

Size has *always* mattered in science fiction, the very, very small invoking the sense of wonder of the techno-sublime just as certainly as does the very, very large. Whatever the cutting edge science behind SF's current fascination with nanotechnology, the metaphorical appeal of the very, very small has been long established. As science fiction codified its modern protocols early in the twentieth century, its stories imagined negotiating vaster and vaster distances, covering more and more time, and even exploring tinier and tinier worlds. As one branch of proto science fiction narratives took travelers to the ends of the Earth, to the Moon, to Mars and other planets, and finally to the stars, racking up ever-increasing mileage, another took travelers inside the Earth, inside drops of water and diamonds, and finally inside the atom, positing and exploring worlds within worlds.

As Brian Stableford explains in his entry on "Great and Small" in *The Encyclopedia of Science Fiction*, different perspectives provided by radical shifts of scale have been central to a long line of satires of which *Gulliver's Travels* is exemplary, and he notes that "shrinking human beings to insect-size in order that they may observe the small-scale wonders of the natural world is common in didactic sf," with Fitz-James O'Brien's "The Diamond Lens" (1858) laying claim to being "the first scientific romance of the microcosm."[1] O'Brien's story foregrounds the science of microscopy, as a stolen diamond makes possible the creation of a perfect lens through which the narrator discovers and falls in love with the beautiful woman Animula, who lives in a world contained within a drop of water.[2] Anticipating both the signature substance of nanotechnology (diamond) and the water drop world of

James Blish's classic "Surface Tension," this story, suggests Thomas Clareson, led the pulp SF writer Ray Cummings to his even tinier character, Lylda, and even tinier "realms of smallness" in "The Girl in the Golden Atom" (1919). While numerous other examples exist of SF's early fascination with the miniature, the microscopic, and even the sub-atomic, two pre-Drexlerian SF stories in particular point toward the construction of nanotechnology in contemporary SF literature: Theodore Sturgeon's "Microcosmic God" and James Blish's "Surface Tension."

"Microcosmic God," first published in 1941, brings together a number of the originary fantasies and/or agendas central to science fiction. Its ostensible protagonist, James Kidder, an expert in biochemistry, creates life, albeit of an extremely miniature form—in the tradition of Victor Frankenstein, although in the clearly American style of the Edisonade. He rapidly evolves his created and confined race of miniature "Neoterics" through a brutal regimen of imposed natural selection, rapidly getting them to the point where their collective brainpower solves problems much faster and more spectacularly than could even a human genius such as Kidder himself. Indeed, SF's preoccupation with genius and endless quest for more brainpower drives this story, as Kidder creates the Neoterics as the only way he can see to speed up intellectual evolution (much the same idea Greg Bear will explore some forty-four years later in *Blood Music*).

While the story ends with Kidder protected from a bothersome outside world by a Neoteric-generated protective dome that encloses the island of their manipulative god, the question remains about what might happen when that god dies and the Neoterics turn their attention to the world outside that dome.[3] Apart from its obvious resonance with the myths of Frankenstein and Edison, "Microcosmic God" anticipates many future SF themes, including several mainstay concerns of nanotechnology-centered stories. The tiny, artificially evolved Neoterics work together—a form of distributed consciousness—to bring about radical changes that transform society, while offering just the vaguest hint of the havoc they might ultimately wreak should they escape the control of their creator. The three-inch Neoterics are not microscopic, much less nano-sized, but Sturgeon's story eerily anticipates the transformative vision that Eric Drexler would codify in *Engines of Creation* and several of the central concerns of nanotechnology narratives in contemporary SF (Sturgeon 1971).

"Microcosmic God" was selected in a vote of the members of the Science Fiction Writers of America (now Science Fiction and Fantasy Writers of America) for inclusion in the 1970 *Science Fiction Hall of Fame* volume of "The Greatest Science Fiction Stories of All Time." Another selection for inclusion in that canonical volume was James Blish's "Surface Tension" (1952), which also features forced evolution but reduces the scale of the characters to the microscopic. "Surface Tension" imagines far-future descendents of space explorers marooned on a distant planet who engineer the survival of humans in an otherwise inhospitable setting by having them evolve into sentient water-breathing beings smaller than a paramecium. These beings, dimly connected to their ancient past by myth, vague racial memory, and enigmatic micro-engraved metal plates left by their ancestors for them to puzzle out, mistake the puddle of water in which they live for the world and are prevented by surface tension from penetrating the barrier of their "sky." Driven to explore, however, and tantalized by the enigmatic metal plate messages from their ancient normal-sized ancestors, they heroically and ingeniously solve the barrier problem of surface tension and discover an adjoining puddle, a new cosmos, with hints of even further worlds having to do with something in their memory called "stars" (Blish 1971).

As was the case in "Microcosmic God," this story implies that small scale does not diminish significance or potential and that evolution can lead to the solving of any problem, the breaching of any barrier. As was also the case with "Microcosmic God," this story incorporates a number of powerful metaphors that will assume even greater significance in nanotechnology narratives. If I am correct in seeing in "Surface Tension" signs of some of the central fantasies of contemporary nanotechnology narratives, there may be an irony in the fact that this story has also been anthologized in David Hartwell's and Kathryn Cramer's 1994 anthology *The Ascent of Wonder: The Evolution of Hard SF*. The irony comes not from its inclusion in this canonical compendium of hard SF, but from the fact that it may well be the *only* story in this collection, published some eight years after Drexler's *Engines of Creation*, to anticipate nanotechnology. The problem that no doubt faced anthologists Hartwell and Cramer in their effort to codify and celebrate the protocols of hard SF was simply that nanotechnology, no matter how rigorously explained or extrapolated *feels* more like magic than like science.

Nanotechnology Narratives: The Vingeian Divide

"I think it's fair to call this event a singularity. . . . It is a point where our old models must be discarded and a new reality rules. As we move closer to this point, it will loom vaster and vaster over human affairs till the notion becomes a commonplace. Yet when it finally happens it may still be a great surprise and a greater unknown"
  ◎ Vernor Vinge, "The Coming Technological Singularity"

"Nanotech makes for good metaphors and for interesting predictions, and wild speculation, and possible terror—all good for science fiction."
  ◎ Kathleen Ann Goonan, Chicon Interview

Although by my conservative count there already exist two anthologies, a good twenty novels, and countless short stories prominently featuring nanotechnology, there is no pressing critical reason to construct a category of science fiction called "the nanotechnology narrative."[4] I've seen calls for some useful buzzword such as "nanopunk" to help distinguish these narratives, but the range of uses to which SF writers have put the concept of nanotechnology hardly lends itself to the idea of even a loosely affiliated movement (like cyberpunk), although nanotech narratives may offer emblems of hyper-embodiment that starkly oppose the emblems of virtuality that cyberpunk codified. And it's hard to imagine a more unlikely assembly of connotations than those that collide when the deliriously utopian-charged "nano" is joined to "punk."

Nevertheless, some clarification of what I mean by "a nanotechnology narrative" will be useful to the discussion that follows. And it may be helpful to divide nanotech narratives into those that envision the transformative power of this technology leading us to some kind of Vingean singularity and those that don't. More about this later. While the term "nanotechnology narrative" is pretty obvious insofar as it suggests stories built from the speculations of Richard Feynman and Eric Drexler, I hope to shift attention from the scientific concepts involved in nanotechnology to the ideological dreams, wishes, and agendas informed and inspired by those scientific concepts in contemporary SF.

Simply put, my thesis is that narratives rising from and/or resting on the concepts of nanotechnology inherently invoke so many central agendas in science fiction that these stories offer particularly useful self-reflexive insights into the nature and appeal of science fiction thinking, a term I've explored at some length in my *Science Fiction After 1900*. My goal in this brief essay is to suggest some of the reasons why nanotechnology narratives strike me as much more than just another set of SF speculations about the impact of a new technology. A secondary goal is to suggest some of the lines along which further study of nanotechnology narratives might develop and to survey the most important writers and texts in this burgeoning field.

Nanotechnology narratives seem to me to offer contemporary SF opportunities to pursue some of SF's most dearly held fantasies with a new claim, if not to hard science rigor, at least to hard science pretense. Within the broad categorization of pre- and post-singularity stories, nanotechnology narratives further fall along a continuum of extrapolative rigor that ranges from the highly technical writing of Wil McCarthy to extremes of whimsical fantasy. While it is easy to criticize the numerous nanotechnology narratives that unrigorously attribute magical properties to nanotech (Ian Watson's *Nanoware Time* comes to mind), it's hard to resist thinking that any technology is not magical that envisions dissassembling any and all matter by breaking it down into constituent molecules and atoms and reassembling those molecules and atoms into an unlimited range of substances, objects, and even human beings.

Wil McCarthy attempted to discourage the easy equation of nanotech with magic in an essay in the fall 2001 *Science Fiction and Fantasy Writers of America Bulletin*. The problem, McCarthy notes in "Nanotechnology: Abuses Of and Replacements For," is that "SF too often falls back on nanotechnology as an almost magical plot device, capable not only of changing texture or shape, but of rapidly building permanent structures and adding exponentially to its own substances."[5] Reminding his readers that nanomachines will have specialized needs for energy as well as for material and that they will best or only constructively function within the confines of climate-controlled reaction vessels, McCarthy also scotches the fantasy of nano-alchemy: "Oh, and transmutation of elements is right out; if your nanocritters are building a city of gold, they'd better have a city's worth of gold on hand."[6] Insisting on the unavoidable limits in speed and method of nanotech transformations, McCarthy ruefully observes: "This fact is so rarely (or poorly) reflected in science fiction that the distinction between technology and magic becomes all but moot."[7]

Of course, McCarthy's own most explicit nanotechnology narrative, *Bloom*, does not meet all of his own criteria for nano-rigor, as it posits a runaway nanotechnology life form, the Mycra, which or who has not only taken over the inner solar system, wresting the planets from human inhabitants, but also has apparently evolved a higher-level sentience. In his more recent novels, *The Wellstone* (2003) and *The Collapsium* (2000), McCarthy has shifted his focus from Drexlerian nanotechnology to the somewhat larger and less fantastic technology of "programmable matter." It is programmable matter that explains the Vingean singularities of the futures in these novels, and McCarthy's interest in the topic has led not only to his nonfiction book, *Hacking Matter* (2003), but also to his collaboration in a patent application for the "wellstone" his science fiction imagines. Discussions and extrapolations of programmable matter might be thought of as "nanotech-lite," offering more immediately achievable results tied to research focused on "utility fog" and "quantum dots." Metaphorically and symbolically, however, distinctions between nanotech and programmable matter narratives are probably too technical to be worth drawing.

The temptation to blur nanotech with magic has also been examined by Graham P. Collins, who observes that in contemporary SF "nanotechnology often becomes a means to accomplish anything within the realm of the imagination, while conveniently ignoring the constraints of physical laws."[8] Interestingly enough, Collins's comments appear in a *Scientific American* article surveying uses of nanotechnology in science fiction, yet another sign of the hazy line between science and fiction where nanotechnology is concerned. On the other hand, Collins also acknowledges a countervailing phenomenon in which science fictional nanotech narratives "reveal some of the actual technical challenges that molecular nanotechnologists might confront if they ever were to execute their designs for real-world nanobots."[9]

My discussion of the various relationships of nanotechnology narratives to science fiction literature will be guided by several loosely connected propositions, the first being that because nanotechnology itself remains largely science fictional, a reciprocal relationship or feedback loop exists between its ostensibly nonfictional and its overtly science fictional narratives. Such a reciprocal relationship is certainly not unique in the history of science fiction, and it is one of the ways in which the efforts of science fiction writers and scientists have been inspired and focused. For science fiction writers, this means a specific sense of mission or agenda above and beyond the epistemological agendas of science fiction in general. Space exploration is one of the best examples of this phenomenon, with robotics and computer science being other obvious instances. What may add significance to the nanotechnology feedback loop is the relative closeness of the visions of the science and the science fiction. University of Washington physics professor and science fiction writer John Cramer explains:

> A foreseen major revolution is perhaps a unique circumstance in human history: a major revolution that is going to have a profound effect on our society, on the way we do things and the way we build things, and that is anticipated long in advance. Its arrival, its impact, and its problems have been anticipated well before the technology is at hand. This did not happen with the industrial revolution, the nuclear age, the space age, or the computer revolution. I cannot think of another example in the history of technology in which the societal impact of a technology has been predicted early enough to allow thorough and coherent thinking and planning. We have time to consider, to steer development, to devise solutions.[10]

A second proposition I will briefly mention is that nanotechnology narratives are logical successors to virtual reality narratives in contemporary SF, which they are in the process of displacing. This is happening in great part, as the discussion of my third proposition will explain, because nanotechnology narratives offer what may be SF's greatest promise of change, and change is at the teleological heart of science fiction thinking. If change is the epistemological and teleological engine that drives science fiction, one of the traditional narrative formulas of SF that has most effectively privileged change is the First Contact story, which I've suggested elsewhere is also of special importance to science fiction because it serves as a self-reflexive metaphor for the experience of reading SF (Landon 2002). My fourth proposition is that part of the allure of nanotechnology narratives is that they offer new takes on the First Contact formula. While too broad a concern to be reduced to a narrative formula, science fiction's fascination with and valorization of genius or brainpower or better thinking leads me to my fifth proposition—that nanotechnology narratives hold special allure because they offer new prospects for better thinking. Finally, I will consider the transcendent or numinous turn taken by a number of important nanotechnology narratives, the prospect of technologized transcendence that appeals to another longstanding dream near the heart of science fiction thinking.

Science, Magic, and Fiction: The Ontological Fuzziness of Nanotechnology Narratives

Before turning to the consideration of a number of nanotechnology narratives clearly shelved in the publishing category of science fiction, it should be noted that not since the glory days of NASA directed space exploration has such a strong reciprocal relationship or feedback loop existed between a science or technology and SF. While scanning tunneling microscopes, atomic force microscopes, and the celebrated IBM logo constructed of individual xenon atoms are but three examples of the already arriving reality of nanotechnology, most of this field remains, at least for the lay public, confined to narrative. Drexler's *Engines of Creation* (1986) opens with a Marvin Minsky "Foreword" invoking the vision of science fiction writers, then turns the vision mission over to Drexler, who incorporates just about every science fiction dream short of FTL travel into his narrative of a totally transformed future. Indeed, it is worth noting that late in this promise-packed book Drexler includes a short section titled "Other Science Fiction Dreams" (emphasis mine), for the SF-primed readers who might need reassurance that telepathy, starships, artificial intelligence, and space settlements may *also* be promoted by nanotechnology, not to mention body modification that "will allow people to change their bodies in ways that range from the trivial to the amazing to the bizarre." "Some people may shed human form," Drexler opines, "as a caterpillar transforms itself to take to the air; others may bring plain humanity to a new perfection."[11] I call attention to passages such as this one not to question the promise or power of nanotechnology, but simply to note that its famous central text reads in many places like science fiction and proudly acknowledges that similarity.

Conversely, more than one commentator has pointed out that only science fiction can begin to cross the cultural chasm that nanotechnology will leave in its wake as it transforms reality in the manner of a Vingean Singularity, itself a concept shared by science and science fiction. In a brief "Introduction" written for Elton Elliott's *Nanodreams,* mostly a science fiction anthology with a couple of nonfiction pieces thrown in, Drexler warns that even some science fiction writers have complained to him that they "can't see how to write realistic stories set in a world with advanced technology," so great is its magical-seeming potential.[12] The great divide in nanotechnology narratives is between those writers who see nanotechnology pushing the future beyond a Vingean singularity, transforming basic assumptions we now hold about the nature of reality, and writers who see nanotechnology adding to the future's inventory of technological marvels, serving more to tweak than to transform human life. For the remainder of this essay, I'll term stories in the former category "post-singularity narratives" and those in the latter "pre-singularity narratives."

Michael Crichton's best-selling *Prey* (2002) is a good example of a pre-singularity nanotechnology narrative, as its plot limits itself to dramatizing the danger of runaway nanodevices in terms of their constructing ruthless pseudo-humans, who, it turns out, act very much like ruthless humans. Greg Bear's *Blood Music* (1985), while not literally an exploration of nanotechnology, is the hands-down exemplum of the post-singularity nanotechnology narrative, as his runaway "noocytes" completely engulf the world, creating a utopian-like vast collective consciousness that sounds a good bit like the standard Christian construction of heaven. While I wouldn't go as far as does Elton Elliott when he accuses pre-singularity nanotech writers of a failure of vision and nerve—an ironic instance of future shock[13] —I will not devote much attention to pre-singularity narratives because they do not seem to engage the significant and almost-certain cultural implications of Drexler's projected nanotechnology revolution. There is no shortage of post-singularity nanotech narratives, as Greg Bear, Kathleen Ann Goonan, Wil McCarthy, and Linda Nagata have alone contributed over a dozen novels to this category, in the process wrestling with both the potential problems and opportunities of nanotech.

The Unsimulation of Cyberspace

> VR always seemed to me to be an interesting dead end—"It was all a Dream."
> ◎ Greg Bear (personal communication)

I'm not sure when we will reach the point at which instances of the prefix "nano-" outnumber instances of the prefix "cyber-" in science fiction, but I have no doubt that we are approaching such a tipping point. Nearly twenty years after William Gibson fascinated us with the concept of cyberspace, many of us spend a good bit of time "there" every day, searching the Web, sending e-mails, playing increasingly realistic computer-animated electronic games. And cyberspace turns out to be much less dramatic and much less transformative than *Neoromancer* suggested it might be, even if its impact on our lives has proved more banally pervasive. Virtuality, perhaps the driving metaphor of cyberpunk and one of the central concerns of the critical discourses of postmodernism, remains an important part of contemporary technoculture, but "virtual reality" as a technology has lost more than a little of its frisson as we have realized that its liberating power fantasies were entertaining, but neither sustaining nor sustainable.[14] The impression of being able to "edit" reality falls far short of the dream of actually doing so. Twenty years after Jaron Lanier was touting the liberatory implications of virtual reality, he's now talking about nanotechnology.

Cyberpunk surfed the technological waves of computer simulation and artificial intelligence, but the movement's most astute and accurate assumption may have been that the future, if not used up, will continue to resemble the present, albeit with neater gadgets. Cyberpunk offered its characters unfettered freedom from the body and from material reality itself as long as they remained in the programmable and easily editable "world" of cyberspace. Nanotechnology narratives offer to make good on the empowering fantasies of cyberspace, pointing to a future in which programmable computer simulations can be realized in the material world and in the physical human body. "With full-scale molecular nanotech," suggests Graham P. Collins, "it is not just *virtual* reality that is programmable."[15]

The supplanting of virtuality in favor of substantiality, of simulation in favor of action in the real world, makes persuasive sense in fictional worlds transformed by nanotechnology. Many, if not most, nanotechnology narratives relegate VR to use in modeling nanotech-transformed reality, turning cyberpunk's emblematic setting into little more than a research tool in nano-inspired settings where almost any material that can be simulated can be created. Frequently, nanotech narratives even comment on this supplanting or development, as is the case in Tony Daniel's Metaplanetary, where nanotech devices called "grist" have been so widely disseminated that they are omnipresent in almost all inhabited physical environments:

The similarities between this interconnected grist and virtual reality of computing entities had long been seen. But there was a visceral, physical quality to the "gristweb." It came to be viewed as the fusion of actuality and virtuality.[16]

A similar "fusion of actuality and virtuality" pervades Kathleen Ann Goonan's "Nanotech Quartet" novels, most notably Queen City Jazz (1996) where Cincinnati, a "Flower City" utterly transformed by nanotechnology, makes possible the compulsion of inhabitants to re-enact settings and stories from literature and jazz history. In a world where nanotechnology makes possible the disassembly and reassembly of matter, virtual reality, tied both metaphorically and metonymically to the promise of computers, gives way to transformable reality, tied both metaphorically and metonymically to the much "wetter" promise of nanotech, not limited to but much more strongly associated with biology. And the VR that remains in nanotechnology narratives tends to be relegated to backseat status. "There are only two industries," says a wise character in Neal Stephenson's The Diamond Age: "There is the industry of things, and the industry of entertainment. The industry of things comes first. It keeps us alive."[17]

No better emblematic testament to the shift from the paradigms of virtuality to those of nano-tech—from the industry of entertainment to the industry of things—can be found than the rejection of "visual ideation," a kind of enhanced virtual reality, in Wil McCarthy's Bloom. There, McCarthy's journalist protagonist, John Strasheim, explains his commitment to the real:

Ideation was a habit, like sweets or stimulants or alcohol, not inherently deviant or harmful in and of itself. Useful in the arts and sciences, of course, and practiced by many respectable citizens. And yet most of the Immunity's ideators simply had too much time on their hands, and too little energy. Why change the world, or even yourself, when you could craft or purchase fantasy environments optimized to your taste and habits? Illegal spec mods aside, the eyes and ears could absorb a great many pleasurable stimuli—not so different, really, from listening to music or going out to the theater or flashing down the occasional VR drama. The temptation was an entirely natural one, and suspect for precisely that reason. Yes, I had done it from time to time, but not often. We had a society to run, now, didn't we?[18]

Assembling Change

Present technologies rely on metal machine systems; future technologies will rely on molecular machine systems. Bulky metal machines squat in our factories and make almost everything we own. Cars and many other products are themselves metal machines, or similarly crude devices of plastic, ceramic, and silicon. These technologies will vanish, probably in the early decades of the next century. What does this mean for SF? Many tales of the future have filled whole galaxies and eons with people using similar machines, much as Greek mythology filled the cosmos with gods riding chariots rather than jets, and wielding arrows rather than guns. Like stone tools, then bronze, then iron, our present technology is on its way to being a historical curiosity. Like stone-age, bronze-age, and iron-age visions of the future, today's SF. . . well, some of SF's classics have already survived futurological obsolescence.

◉ Eric Drexler, "From Nanodreams to Realities"

"The thing is—nanotech really IS coming, and I think that there really WILL be vast changes because of it. And so. . . I write."

◉ Kathleen Ann Goonan, Chicon Interview

Science fiction is the literature of change. More precisely, science fiction is the kind of literature that explicitly and self-consciously takes change as its subject and its teleology. While this view of SF is neither original nor controversial, I've devoted quite a bit of print to elaborating it in order to remind new readers that SF, at its essential core, uses science and technology in the service of agendas more than as subjects. Indeed, for me, nothing in science fiction better indicates its profound commitment to insisting on the primacy of change than does the secular credo of "Earthseed," developed by Octavia Butler's protagonist Lauren Olamina in *Parable of the Sower* (1993). "Earthseed" holds that, ultimately, "God is Change," with Change constructed as the only lasting truth in, and organizing principle of, existence.[19]

If such a credo comes close to specifying the central tenet of science fiction as a belief system, post-singularity nanotechnology narratives may be the apotheosis of SF. This most radical technology for constructive change practically demands science fictional extrapolation and valorization.

Second Tries at First Contact

Ian R. MacLeod's "New Light on the Drake Equation" offers a poignant look at a vastly changed late twenty-first century earth in which independent researcher Tom Kelley is one of the last, if not the last, SETI Project true believers (MacLeod 2002). Each night Kelley waits in vain for the message from outer space that hasn't come; even the calculations of the Drake Equation no longer inspire much confidence that it ever will. Not only is Kelley a throwback in his continuing faith in first contact, he is also a throwback in his resistance to using the nanotechnology that has utterly transformed life on Earth. He lives in France but declines to use the nanomolecules in a vial that would specifically enhance his language and cognition centers so that he could understand and speak French. Likewise, he eschews the wings and gills, ornamental green scales, and other body modifications, both utilitarian and ornamental, made possible by readily available nanotechnology. The not-so-subtle point of MacLeod's story is that Kelley is so firmly invested in a fairly narrow and traditionally science fictional notion of first contact with aliens from the stars that he has not noticed how alienated he has become from his fellow humans, nor how alien they have become through the transformations made possible by nanotechnology. In a sense, MacLeod's story offers us the example of a stealth post-singularity nanotechnology narrative: the singularity has occurred, but at least the protagonist didn't notice.

More important, this story calls attention to the ways in which nanotechnology is being used by science fiction writers in conjunction with the traditional first contact formula. While Drexler himself, much like MacLeod, was more interested in *Engines of Creation* in discouraging obsession with the discovery of other intelligent species in the cosmos, the idea of nanotechnology seems to have had quite the opposite effect on SF writers. The extrapolative logic is clear: if nanotech represents for us revolutionary technological advancement, why would it not also figure prominently in the technologies of alien civilizations SF usually constructs as more advanced than ours? This is the central assumption in Kevin J. Anderson's and Doug Beason's *Assemblers of Infinity*, which juxtaposes fledgling nanotech research on Earth with the discovery on the far side of the Moon of a marvelous nanotechnology-constructed artifact, the result of alien efforts to seed space with nanotech devices. Exploring the hypothesis that aliens had broadcast these nanotech devices, one of the Earth researchers in this novel reasons that nanotech machines carrying enormous computing power could be accelerated without harm to nearly the speed of light and would be the obvious messenger and/or message of choice from an advanced alien species:

> Well, maybe nanotechnology is the way they think. Say their society is based on nanotechnology—they wouldn't consider sending massive objects when they can send little probes programmed to build what they want when they get here.[20]

As it turns out, the alien nano-constructed structure on the Moon serves both as a communication device sending information of its contact with humans back to its alien designers and as an elaborate incubator and educational resource for the alien embryos it houses, themselves created from genetic information carried by nanotech machines.

First contact alien contamination that turns out to be incredibly beneficial to the poor and powerless of Africa while predictably threatening to all existing world political and economic orders drives Ian McDonald's "Chaga" series, including the novel *Evolution's Shore* (McDonald 1995) and the short stories "Recording Angel" (McDonald 1998) and "Tendeleo's Story" (McDonald 2001). As is frequently the case in nanotechnology narratives, McDonald turns the contagion and/or alien invasion formulae on their head, positing alien nanotech as salvation for third world problems.

Nanotechnology narratives offer SF writers a range of variations on the first contact formula: they can construct nanotechnology as the technology that enables us to make first contact with aliens, that enables aliens to make first contact with us (also marking alien technological advancement), or they can ascribe to nanotechnology some form of emergent consciousness and intelligence, with nano machines or lifeforms themselves constituting the previously unknown species we encounter for the first time. Crichton's *Prey* offers a hint of this latter formula, although its author eschews big questions about the nature of our interaction with a new intelligent life-form in favor of a conventional thriller that is mostly a domestic drama (Crichton 2002). Much more provocative interrogations of this possibility are Wil McCarthy's *Bloom* (McCarthy 1998), Stephen Baxter's "The Logic Pool" (Baxter 1998), and Paul Di Filippo's "Up the Lazy River" (Di Filippo 1998).

For a Nano Paideuma

"Frobenius uses the term Paideuma for the tangle or complex of the inrooted ideas of any period. . . . I shall use Paideuma for the gristly roots of ideas that are in action."
  ◎ Ezra Pound

"To prepare for the assembler breakthrough, society must learn to learn faster."
  ◎ K. Eric Drexler, *Engines of Creation*

Arguing for a new approach to providing "a minimum for a decent education in our time," and aiming for an education that would reconceive culture, Ezra Pound borrowed the term "paideuma" from Leo Frobenius and called for a New Paideuma.[21] Pound liked the term for its vitality, its expansiveness, and its potential for indicating sweeping change in educational assumptions, new learning that would lead to a new civilization. To my knowledge, no nanotechnology narrative has yet followed Pound in using this word, but a fascinating constant in nanotechnology narratives is a focus on education and assumptions that nanotech can revolutionize learning just as surely as it can transform materials. Indeed, running through nanotechnology narratives in SF is an even more fascinating complex of associations of nanotechnology, book technology, learning, and the promise of technology-enhanced intelligence. This complex of associations brings together the theorizing of such different thinkers as Richard P. Feynman and Susan Stewart with fiction writers such as Neal Stephenson and Kathleen Ann Goonan.

Many trace the origin of contemporary interest in nanotechnology to Feynman's famous talk, "There's Plenty of Room at the Bottom," delivered in 1959 at the annual meeting of the American Physical Society.[22] Feynman works down to the scale of his ultimate subject, nanotechnology, by first considering miniaturization technologies that would allow writing the Lord's Prayer on the head of a pin, and then ups the miniaturization ante by claiming that there "is no question that there is enough room on the head of a pin to put all of the *Encyclopedia Britannica.*" His final scalar change envisions bringing together in one nano-scale work "all of the information which all of mankind has ever recorded in books," a body of work Feynman estimates at some 24 million volumes. Feynman's linking of nanotechnology and book technology gave him a powerful emblem for thinking about the dimensional potential of nanotech, but it also (almost certainly without Feynman's knowledge) linked the dream of nanotech to the nineteenth century's fascination with miniature books, a fascination brilliantly theorized by Susan Stewart in her *On Longing: Narratives of the Miniature, the Gigantic, the Souvenir, the Collection.* I don't have the space in this essay (there's an irony there!) to trace the connections between Stewart's work on miniaturization and the assumptions of nanotechnology theory and nanotechnology narratives, but her thesis that nineteenth-century miniature books were intended to "connect the book to the body" and, indeed, to make a "digestible book" resonates with and offers valuable insights into the nano paideuma dreams of SF nanotechnology narratives.[23] And nowhere do these resonances and insights better come into play than when they are brought to the reading of what may well be the most accomplished of all nanotechnology narratives, Neal Stephenson's *The Diamond Age.*

Subtitled "or, A Young Lady's Illustrated Primer," *The Diamond Age* constructs a nanotech-transformed future in which the most significant of all nanotech-driven innovations may be the ultimate electronic text, a kind of pedagogical Turing Machine that by itself provides a nearly complete interactive education. Intended only for the eyes and mind of an elite technocrat's daughter, in hope that it will teach her to think for herself, free of the hegemonic limitations of conventional education, this *Young Lady's Illustrated Primer* is itself subtitled "a Propaedeutic Enchiridion in which is told the tale of Princess Nell and her various friends, kin, associates, &c." The problem is that the technocrat's daughter for whom this hyper-encyclopedic teaching device was intended is named Elizabeth and Nell is an urchin to whom her street thug brother gives an illicit copy of the primer. The book then "bonds" with Nell, reshaping its lessons to her surroundings and personality and ultimately producing a young woman whose nanotech enabled education prepares her to change the world.

While Stephenson's *Young Lady's Illustrated Primer* seems less like a familiar product of nanotechnology than an artificial intelligence that has hypertext access to all information, an amazing amount of knowledge, and quite a bit of wisdom than a familiar product of nanotechnology, nanotech makes its virtual reality pedagogical scenarios possible as it is the ultimate example of "smart page" technology. Readers are left to infer much of the way Stephenson's primer works, but a very good call for something like the primer can be found in Drexler's *Engines of Creation,* particularly in his chapter "The Network of Knowledge," whose subsections include a discussion of "Magic Paper Made Real." There, Drexler claims that nanotechnology will make possible a "book-sized object" that "will be able to hold a hypertext system containing images of every page in every book in the world, storied in fast, molecular-tape memory."[24] Recognizing the advantages of locating information in engaging narratives, perhaps the basic premise underlying Stephenson's primer, Drexler goes on to suggest that by offering miniature "three-dimensional television so real that the screen seems like a window into another world," nanotechnology will also give a future learner "vivid art forms and fantasy worlds far more absorbing than any book, game, or movie."[25]

Stephenson has most thoroughly taken Drexler's vision of a nano paideuma to heart, but few writers of nanotechnology narratives extrapolate this technology without envisioning its dramatic implications for education and enhanced intelligence. Kevin Anderson and Doug Beason don't go as far down the road of pedagogical prescription in their *Assemblers of Infinity,* but the alien embryos constituted by nanotechnology in their ingenious first contact story must be completely raised and tutored by nanotech-built devices, a process one of their characters incongruously likens to Tarzan's being raised by the apes (Anderson and Beason 1993). The eponymous Bohr Maker in Linda Nagata's novel offers us another glimpse of a nano device that bestows extraordinary expertise on its host (Nagata 1995).

So prevalent is this pattern in SF nanotechnology narratives that Gregory Benford singles it out as one of the primary abuses of science we find in extrapolations of nanotech:

> We see stories about quantum, biomolecular brains for space robots to conquer the stars. Or accelerated education of our young by nanorobots which coast through their brains, bringing encyclopedias of knowledge disguised in a single mouthful of Koolaid.[26]

Benford may have had stories in mind such as Michael F. Flynn's "Remember'd Kisses," with its nanotech version of the Pygmalion myth (Flynn 1998). Flynn's protagonist uses nanotechnology to remake the mind and personality of a homeless bag lady, giving her many of the memories and much of the mental ability of his dead wife. Anthologized in the Dann and Dozois *Nanotech,* "Remember'd Kisses" also appears in Flynn's apparent fix-up novel, *The Nanotech Chronicles,* where the nano paideuma becomes part of the narrative framework as its narrator is himself an educator, instructing his elementary school students in the "history" of nanotechnology (Flynn 1991).

Concerned less with education than with nanotech brainwashing or brain-altering are stories such as Greg Egan's "Axiomatic" (Egan 1998) and Paul Di Filippo's "Any Major Dude" (Di Filippo 1998), both of which envision nanotech altering or shaping of free will. Underlying the action of Greg Bear's *Queen of Angels* (Bear 1990) and *Slant* (Bear 1991) is the assumption that nanomachines can repair or adjust unbalanced or unhealthy minds through changing the structures of the brain with nano therapy. Similar assumptions about the ability of nano devices to protect, *or even to supplant parts of*, the mind can be found in Alastaire Reynolds's "The Great Wall of Mars" (Reynolds 2001) and "Glacial" (Reynolds 2002). In somewhat similar, but more overtly ominous, fashion, Kathleen Ann Goonan, Tony Daniel, and Paul Di Filipo all envision nano-weapons that can spread a particular ideology or compulsion, as well as being able to deprive the brain of specific conceptual and linguistic abilities. Goonan imagines Information Wars in which the primary weapons are "nanotech-infused clouds which carry viruses of thought," and she specifies in her Chicon interview that these viruses will work "like, perhaps, religion, or any kind of political dogma."[27] It is also worth noting that all of Goonan's Nanotech Quartet" *(Queen City Jazz, Crescent City Rhapsody, Mississippi Blues, Light Music)* novels construct nanotechnology primarily in terms of its ability to carry and hypertext information.

Regardless of whether these stories construct nanotech's ability to change minds positively or negatively, the promise of dramatic technological impact on thinking adds immeasurably to the allure of nanotechnology for science fiction writers. Not surprisingly for a mode of fiction fascinated by brainpower and genius, science fiction returns again and again to stories in which some new system of thought or technology leads to a dramatic increase in intelligence. Prometheus, Faust, Frankenstein, and a host of characters and works not generally associated with science fiction remind us of the universality of this concern, but science fiction has uniquely attempted to technologize this dream. John Huntington has identified and brilliantly explored this theme in his *Rationalizing Genius: Ideological Strategies in the Classic American Short Story,* a pioneering work of SF scholarship whose insights and generalizations hold true for far more than the twenty six *Science Fiction Hall of Fame Stories* Huntington took as his sample (Huntington 1989).

Indeed, the rationalization and valorization of genius continues to provide one of the clearest ideological agendas in contemporary SF, frequently displacing the focus on intellect from humans to aliens, computers, and artificial intelligences, but always obsessed with ways of getting smarter. It is also not surprising that such an agenda would lead SF narratives again and again to stories centered on the intelligence enhancement and super-education of children in classic stories such as Lewis Padgett's "Mimsy Were the Borogoves" (1943) (Padgett 1971) and Theodore Sturgeon's "Tandy's Story" (1961) (Sturgeon 1971). The explicit process of education, often with accompanying pedagogical theories, drives the narratives of a large number of science fiction's most celebrated novels and stories. Certainly, Stephenson's *The Diamond Age* adds brilliantly to this tradition. Against such a background, the promise of the nano paideuma takes on particular significance and appeal, especially since two of the four possibilities considered by Vinge as likely causes for a technological singularity involve the "superhuman" development of human intellect.

Gray Goo and/or Transcendence?

What is "the technological sublime," and why do we swoon for it? Why do we fall for this silly-sounding thing? It is the projection of a spiritual need for transcendence onto mechanical hardware. A sublime thing inspires awe and wonder, it ruptures the everyday. The sublime is a liberating spectacle that lifts the human spirit to plateaus of high imagination. The most alluring and attractive way to deal with technology is to hype it as something divorced from the normal routines of life.

⊚ Bruce Sterling, *Tomorrow Now*

"Are you still a human, Michael"

⊚ *Blood Music,* by Greg Bear

The gray goo threat makes one thing perfectly clear: we cannot afford certain kinds of accidents with replicating assemblers.

⊚ K. Eric Drexler, *Engines of Creation*

In his post-singularity story "We Were Out of Our Minds with Joy," David Marusek pauses long enough amid descriptions of nanotechnological life extension and reassembled bodies to remind us that even miraculous technology can have mundane, even banal, applications. Marusek's protagonist is a nano-artist whose most successful creations are novelty gift wraps, including video wrapping paper that displays moving images and projects sound, wrapping paper that looks like human skin that bleeds and screams when cut, and wrapping paper that is actually a biological orange peel that squirts citrus juice when cut.[28] In the context of the staggering claims and imagined marvels that fill nanotechnology narratives in SF, such a reminder offers a moment of comic relief, at least a small reality check, and an ironic counterpoint to the greatest imagined threat of nanotechnology—gray goo—and its greatest imagined good—some form of transcendent consciousness, a utopian community of thought.

Drexler reminds us that the threat of runaway dissassemblers and replicators is not necessarily tied to a totalized nano environment that is either gray or gooey,[29] but the term is universally recognized in the various discourses of nanotech. The fear is of nano-replicators, whether mechanical or biological, run amok, converting everything they contact into more amok nano-replicators until they are all that remain. Greg Bear perfectly captured this threat in his proto-nanotech masterpiece, *Blood Music*, where biological noocytes take over and completely transform not only the world as we know it, but the basic operations of and assumptions of reality itself (Bear 1985). The other paradigmatic "gray goo" novel is Wil McCarthy's *Bloom*, which features a runaway nano-entity called the Mycora, so insatiable that it has displaced humans from all of the inner planets of the solar system (McCarthy 1998). And Ian McDonald's "Chaga" stories posits an alien sent gray goo that is "eating" and transforming Africa, offering a somewhat harder-science, nanotech-driven version of Ballard's *The Crystal World* (Ballard 1966). Ever the linguistic-innovator, Paul Di Filippo names the sentient gray goo in his "Distributed Mind" URB, short for "Urblastema" and interchangable with "Panplasmodaemonium" (Di Filippo 1996).

Interestingly enough, while all of the above gray goo nanotechnology narratives effectively dramatize the totalizing threat of unchecked nano-replication, all also ultimately yield to the fantasy that this is a good thing, a next stage in evolution, a step toward a higher consciousness. All suggest that humans dissolved into the gray goo can be extruded again from it, although they probably wouldn't want to give up their new-found collective or distributed consciousness. Di Filippo goes so far as to suggest that URB is God, and Bear, one of SF's current hard science masters, makes the noosphere or thought universe that follows hard on the heels of a gray-goo-like transformation of Earth sound very much like the most wishful Christian stereotypes of heaven. It turns out the runaway Mycora in McCarthy's *Bloom* haven't necessarily destroyed the humans it has disassembled, but has "unpacked" them, liberating them from the limited form of solid flesh, combining billions of human consciousnesses in the vast emergent consciousness of the Micosystem.

The alien Chaga in "Tendeleo's Story" offers "different ways of being human,"[30] ways that are clearly meant to be superior to the old ways; "every rule about how we make our things, how we deal with each other, how we lead our lives, [are] all overturned"[31] in a nano-utopia. The world according to Kathleen Ann Goonan's "Nanotech Quartet" novels has been terribly transformed by runaway nanotech (though not exactly by gray goo), but nanotech also continues to offer hope of a utopian "Nawlins," featured in *Crescent City Rhapsody* (Goonan 2000) and then on to new forms of awakened consciousness in *Light Music* (Goonan 2002 b).

The promise that it may lead to forms of transcendence may be the most deeply rooted allure of nanotechnology narratives for science fiction. Some SF critics, most notably Alexei and Cory Panshin in *The World Beyond the Hill* have argued that transcendence is the ultimate goal of science fiction writers.[32] At least so far as nanotechnology narratives are concerned, they may well be proved right. "Transcendence, and the possibility of transcendence," says Kathleen Ann Goonan in her *Locus* interview, "is the aspect of being human—and of science fiction—that keeps me going,"[33] and nanotechnology has obviously provided her with a technological means toward that imaginative end.

"Transenlightenment" becomes possible through nanomachines that allow the merging of human minds in the Conjoiner stories by Alastair Reynolds, particularly "The Great Wall of Mars," where a character attempts to explain the phenomenon:

> Afterward, when Clavain tried to imagine how he might describe it, he found that words were never going to be adequate for the task. And that was no surprise: evolution had shaped language to convey many concepts, but going from a single to a networked topology of self was not among them. But if he could not convey the core of the experience, he could at least skirt its essence with metaphor. It was like standing on the shore of an ocean, being engulfed by a wave taller than himself. For a moment he sought the surface; tried to keep the water from his lungs. But there happened not to be a surface. What had consumed him extended infinitely in all directions. He could only submit to it. Yet as the moments slipped by it turned from something terrifying in its unfamiliarity to something he could begin to adapt to; something that even began in the tiniest way to seem comforting.[34]

Indeed, in yet another reminder of the fuzzy interface between fact and fiction where nanotechnology is concerned, the Transhumanist movement points to some nanotechnology narratives to illustrate its goal of affirming "the possibility and desirability of fundamentally improving the human condition through applied reason, especially by developing and making widely available technologies to eliminate aging and greatly enhance human intellectual, physical, and psychological capacities."[35]

In some ways, the dreams of transcendence in nanotechnology narratives can be seen as an updating of the longstanding SF dream of telepathy, and the valorization of nano-enabled collective consciousness can be seen as an interesting rehabilitation of the concept of the "hive mind," long one of SF's greatest fears. Whatever the thematic agenda, nanotechnology narratives in science fiction allow writers to re-access and recycle some of SF's most dearly held fantasies. Whether transcendence comes in the numinous form of Bear's noosphere and Di Filippo's URB or in the form of enhanced and expanded consciousness found in nanotechnology narratives by Goonan, McCarthy, McDonald, and Reynolds, there is a clear tendency in science fiction to construct nanotech as a singularity-sized step toward enlightenment.

◇◯◇

Future Present: Nanotechnology and the scene of risk
KATE MARSHALL

"You apply mathematics and other disciplines, yes. But in the end you're dealing with a system that's out of control. Hysteria at high speeds, day to day, minute to minute. People in free societies don't have to fear the pathology of the state. We create our own frenzy, our own mass convulsions, driven by thinking machines that we have no final authority over. The frenzy is barely noticeable most of the time."

⊙ Don DeLillo, *Cosmopolis*[1]

"I have seen the future, and it is small."

⊙ Larry Smarr, "Nano Space," in the June 2003 issue of *Wired*[2]

Charles, Prince of Wales, is worried about the future. He is an outspoken campaigner against developments in genetically modified food, high-rise building, and now nanotechnology. His Foundation for the Built Environment is attempting to preserve what he calls the "human scale" in built structures, before urban architecture progresses further upward. In a speech given to the Société de Géographie at la Sorbonne in February 2003, he used the rhetoric of his fight against genetic engineering to argue for architectural limitations: "After all, in this world of instant communications, jet aircraft and the video conference, why do people remain so attached to their village, their town, or their region? Why are so many people drawn to the more 'organic' character of traditional settlements than what might be termed the 'genetically engineered' and soulless developments of our contemporary world?"[3] If the contemporary world, in this instance, has no soul, then the future looming just beyond it is the threatening force against which the prince is trying to set up defenses. He opposes technological development to an older sense of space and environment, and in doing so calls upon his work to limit the biological engineering of "traditional" foodstuffs. So when, in April 2003, he was reported to have asked the Royal Society to examine the "enormous social and environmental risks" of nanotechnology,[4] the combination of scale and science as his concerns cohered with his previous public agenda. By highlighting the potential size of nanotechnology's scope and scale as ubiquitous and global, Prince Charles turns the science of the small into the technology of a vast and runaway future. And rather than promoting a purely reactionary stance, his public statements on genetic engineering, lived space, and nanotechnology are forward-moving, more anticipatory than backlash. He is reacting to the potential future of development over its current achievement. As biological science, space, and technology come together in the figure of incipient sovereignty, the shadow of their uncharted futures reaches back in the form of risk consciousness.

The question of authority when dealing with a technology that is an application of several disciplines is necessarily an open one, and nanotechnology, as the flagship of scientific and technological convergence, is driving concurrent debates in research, politics, and popular fiction that agree to focus on one point: its future trajectory. In the following pages, I will consider how the forecast of the technology feeds back into the present cultural moment. By anticipating nanotechnology's future and projecting images of its spatial form, the novels I will describe construct nanotechnology as a risk technology, one that emerges as a reflexively produced technological threat, and whose present conception is determined by its charted future. I will discuss the social production of the concept of nanotechnology and the stakes for fiction in that matrix, and examine the consequences of convergent technologies for that fiction. I will also locate nanotechnology within the sphere of reflexive modernity, the site of the risk society and its future-orientation. This shifted temporality, and its accompanying realignment of space, leads to a fictionally contemplated sacrifice of the material world for the sake of progress, or the reverse: the sacrifice of progress for the sake of the material world. The novels that take part in this interrogation, primarily Michael Crichton's *Prey* and Michel Houellebecq's *The Elementary Particles,* and also other works that take issue with the developments of nanotechnology, reveal a complicated notion of risk that drives the perception of a technology whose present is located largely in its future.

The social production of nanotechnology

Nanotechnology, as an object of scientific inquiry, a recipient of public and private funding, and the subject of popular fiction, is a social concept, one which cannot be dissociated from the heterogeneous forces pulling at its present and future shape. The triple construction of scientific, econo-political, and social forces in the production of nanotechnology requires an examination of how scientific knowledge addresses itself to cultural positions, and what role the technology's imagined futures have in reinscribing its status as a social concept.

Contemporary phenomena tied to technological development such as the ozone hole tend to resist any single theoretical approach, because the networked nature of their construction functions as a fundamental barrier to interpretation. In the example of the ozone hole, nature, power, interest, and meaning overlap, canceling each other's claims to priority. Bruno Latour asks: "Is it our fault if the networks are *simultaneously real, like nature, narrated, like discourse, and collective, like society?*[5] While it can be said that the ozone hole itself is more of a result of technology than an object of development like nanotechnology, the two operate culturally in very similar ways, as created and mediated *quasi-objects* developing through the collision of political, economic, and cultural forces with the science they both describe and produce.[6] The by-products of technological development exist through multiple definitions because of their networked nature, and nanotechnology, especially because of the indeterminate future that produces its multiple realities in the present, operates similarly.

As an object of growing public debate, nanotechnology receives multiple claims to authority. Michael Crichton's *Prey,* published in 2002 to a predictably large audience, is a prominent example of fiction weighing in as an ethical arbiter of the potential futures of the technology. Crichton prefaces the novel with a footnoted introduction that calls for international controls on nanotechnology-related research and production, but offers up his fictional account as a site of reasonable speculation for the justification of such controls: "But of course, it is always possible that we will not establish controls. Or that someone will manage to create artificial, self-reproducing organisms far sooner than anyone expected. If so, it is difficult to anticipate what the consequences might be. That is the subject of the present novel".[7] Reviews of *Prey,* especially in scientific publications, are quick to dismiss Crichton's vision as "bad science" (see for example, Chris Phoenix's "Don't Let Crichton's *Prey* scare you–the science isn't real," *Nanotechnology Now,* January 2003), but raise the question of who exactly has claims to scientific knowledge when dealing with a socially produced body such as nanotechnology.

The modern society, and correspondingly, the modern rationality, in which nanotechnology situates itself is characterized by displaced claims to scientific knowledge. The ownership of knowledge shifts from the lab to the corporate sponsor, from the governmental committee to the public watchdog. Sociologists Ulrich Beck and Anthony Giddens consider this refusal to locate scientific knowledge resolutely with scientists themselves a symptom of *reflexive modernity,* that is, a condition in which society defines itself by reflecting upon the technological risks it has created. This self-produced risk, a result and initiator of technological development, alters the power structures of scientific knowledge so that development itself is seen as the proper object of regulation outside of its disciplinary boundaries. According to Beck, a loss of cultural consensus on the good (or otherwise) consequences of techno-economic development transfers the power of legitimation of technology to the public forum.[8] Furthermore, the broken monopoly on scientific rationality leads to a kind of "democritization" of science that is in itself self-produced: "In their concern with risks, the natural sciences have involuntarily and invisibly *disempowered themselves somewhat, forced themselves toward democracy".[9]* At the same time that those doing research must take risk into account, the act of doing so extends the domain of knowledge to the public. In reflexive modernity, risk is the double bind that technoscience must inhabit. I will deal more explicitly with nanotechnology's relation to risk later, but its relation to a fear of consequences that often accompanies scientific and technological inquiry within the state of reflexive modernity in part drives the claims to knowledge I am describing here.

Giddens pursues this line of thought to claim that any form of social belief in scientific progress relies on the assumption that human activity, including any technology that can affect the material world, is a social creation, and as a result, nothing in science can be said to be certain or provable in that context: "In the heart of the world of hard science, modernity floats free."[10] The pluralizing operation performed on knowledge can also be attributed to theories of post-modernity and their accompanying dismissal of the *grand recit* and homogenous narratives of knowledge, which Giddens acknowledges and then reinscribes in radicalized, reflexive modernity that analyzes, rather than breaks with, itself.[11]

The question of what sort of modernity nanotechnology inhabits returns to the idea of the "narrative" of science that Latour includes in his characterization. The narratives of nanotechnology have already engaged in a dialogue, as shown with *Prey,* that emphasizes their place in the social production of the concept. They form part of what N. Katherine Hayles calls the "cultural matrix" as a model for examining the forces which guide scientific and intellectual inquiry.[12] Rather than drawing direct causal lines of influence, she says, an appropriate analysis of interdisciplinary exchange would have to consider the constellation of forces that accompany paradigm change.

The matrix surrounding nanotechnology is an accelerated one: so much so that the very definition of what the terms "nanotechnology," and "nanoscience" are is almost lost in overdetermination. It risks becoming what Rem Koolhaas describes as "Junkspace": "like a best-seller–overdetermined and indeterminate at the same time."[13] IBM, which emphasizes its role in the creation of the primary materialities of nanotechnology, including the Scanning Tunneling Microscope that makes manipulation of individual atoms possible, confidently provides definitions on its corporate website (www.ibm.com): "Nanoscience is the discipline of understanding how things work, using commonly accepted experimental and theoretical techniques. Nanotechnology is the application of that scientific knowledge to a particular industry or marketplace." Other definitions, of course, abound, usually revolving around the scalar definition of the nanometer. Scientists, working in the corporate and academic arena alike, are loathe to relinquish epistemological control. Ralph Merkle, associated with the Foresight Institute, complains that overdetermination of the term robs it of meaning within the scientific circle itself: "Many researchers wish to adopt a definition of 'nanotechnology' which includes their own work. An unfortunate consequence of this is that the unqualified term 'nanotechnology' comes to mean very little."[14] But any prospect of "qualifying" such a distributed term automatically becomes subject to its own objections.

While fiction may not qualify any singular definition of nanotechnology or the larger scope of its cultural significance, it does exert influence over the term's quality as an arbiter and producer of cultural forces. A split emerges, for example, between Greg Bear's early sounding board *Blood Music* (1985) and the fall 2003 premier of UPN's television series *Jake 2.0*. *Blood Music* includes the complete biological transformation of North America within its portrayal of the consequences of nanotechnological development, and places much of the blame for this destruction on one scientist's myopic vision: "He may not have had the time, but even allowing him the time, Vergil simply did not think such things through. Brilliant in the creation, slovenly in the consideration of consequences."[15] The text containing these words is in fact a consideration of those consequences, and may be read as a (rather more subtle than Crichton's *Prey)* call to do so, even if the novel itself refuses to pass final judgment on the consequences it imagines. Alternatively, *Jake 2.0* presents nanotechnology as the gateway to a young geek's hero fantasy. "It's like *Spider Man* meets the *Six Billion* (sic) *Dollar Man,"* a PR representative for the show told me–big money meets small technology meets national security. The premise of the show is that a technical support worker for the National Security Agency gets accidentally exposed to "nanites," which give him strength, speed, heightened senses, and an uncanny ability to plug into any computer or satellite on the planet. This updated version of Jake, while uncomfortable with his sudden transformation and its implications, is nonetheless more positive about the power and potential for nanotechnology. And given that Hewlett-Packard has launched a series of television advertisements touting the potential for nanotechnology to produce "a cell phone so small an ant could use it," enhanced human performance is also critical to portrayal of nanotechnological futures.

I am not attempting to take stock of the forecasting capability of fiction, which Marshall McLuhan describes as "the power of arts to anticipate future social and technological developments," but rather the conditions of representation specific to nanotechnology and the consequences of authors' representational choices.[16] Latour's "Research Space" contribution to the June 2003 *Wired* discusses a representational circuit, but one that conspicuously omits fiction:

> The matters of fact of science become matters of concern of politics. As a result, contemporary scientific controversies are emerging in what have been called hybrid forums: one, science, representing nature–here "representation" means accuracy, precision, and

reference—and another, politics, representing society—and here "representation" means faithfulness, election, obedience.[17]

A third consideration of representation must be included in this formula of hybridity if applied to nanotechnology, fictional representation, which here means analysis, entertainment, and relevance. Within this field of representation, the public-as-audience can locate itself within the *portrayed* world,[18] which is one of the most accessible (and accessed) methods for engaging with nanotechnology within the non-specialized public. The conflation of representational methodologies and meanings compares structurally with the converged technologies that create nanotechnology and come together in its fictional representations.

Convergence: "Oboy."

The prospect of biology, information technology, and nanoscience converging drives the speculative aspect of the concept of nanotechnology, and corresponds with a gravitational trend displayed by economical, technological, and social developments since the middle of the twentieth century. This coalescing operation forms the networking principle set forth by Manuel Castells, which incorporates the processes of: "the information technology revolution; the economic crisis of both capitalism and statism, and their subsequent restructuring; and the blooming of cultural social movements, such as libertarianism, human rights, feminism, and environmentalism."[19] These come together and their interactions create, he claims, a new, networked, social structure. The convergence of social and technical constructions, then, feeds into a cultural matrix thematized by convergence, in which nano-technology acquires a central position.

A RAND Institute publication, *The Global Technology Revolution,* focuses on what it deems "synergies" between the converging sciences which meet in nanotechnology. According to the publication, "life in 2015 will be revolutionized by the growing effect of multidisciplinary technology across all dimensions of life: social, economic, political, and personal."[20] A Department of Defense-affiliated research group, RAND approaches the topic of nanotechnology with an eye to the potential power of such convergence. A recent nanotech conference at the University of California, Los Angeles, "Converging Technologies for Improving Human Performance," used the acronym NBIC, for nano, bio, info, and cogno,[21] to produce linguistically the site of convergence (Smarr's *Wired* piece calls it "bioinfonano-tech"). Given that Jake Foley of *Jake 2.0* is described, among other things, as a "supersoldier," it's easy to see how defense interests and improved human performance are themselves subject to the phenomenon of convergence.

In Michel Houellebecq's *The Elementary Particles,* the principal character is a scientist of two disciplines, the combination of which allows him to change radically human life on earth. Michel Djerzinski works in a molecular biology facility, in which he and his supervisor are "probably the only members of the National Scientific Research Center who had studied physics and who understood that once biologists were forced to confront the atomic basis of life, the very foundations of modern biology would be blown away."[22] Here Houellebecq locates the site of convergence in the character. Later, he will locate it within the human cell, but always, it is within life. Crichton, however, brings converged technologies to life into a theater of self-organization and biomechanical autonomy. *Prey* focuses on the danger of combining the replicating properties of bacteria with microtechnology, in which the process of doing so eliminates the possibility of human control. The technology in question is a "swarm" of tiny cameras which are designed to move as a unit to capture an image. In order to solve technical difficulties involved in getting the individual units to move together, the scientists responsible for their development rely on emergent behavior, which is unpredictable and difficult to control: "In a way this was very exciting. For the first time, a program could produce results that absolutely could not be predicted by the programmer. These programs behaved more like living organisms than man-made automatons."[23]

Emergent behavior, for Crichton, is what makes nanotechnology so dangerous. And he is not alone in his view. Bill McKibben, whose 2003 book *Enough: Staying Human in an Engineered Age* claims that the technological achievement of humanity is nearly sufficient to necessitate a reevaluation of whether the risks of continued progress outweigh its benefits, argues that a multidisciplinary technology revolution could mean the destruction of humanity. He cites a *New York Times Book Review* piece by Thomas Pynchon that discusses the prospect of combined technologies: "if our world survives, the next great challenge to watch out for will come—you heard it here first—when the curves of research and development in artificial intelligence, molecular biology, and robotics all converge. Oboy." Pynchon's piece was in 1984, and in 2003, McKibben says, that "oboy" probably refers to nanotechnology. It is not my intention to pass judgment on the likelihood of dystopian claims like McKibben's or Crichton's, but rather to examine the social factors that produce and react to the fear of nanotechnology and its convergent nature. I have already discussed the status of nanotechnology as a social production, and now I would like to deal with the social and scientific relations to nanotechnology's negative image, and what part fiction has to play in its creation and interpretation.

A Risk Technology

What is it about nanotechnology and its representations that make McKibben (who mentions Crichton's *Prey,* Neal Stephenson's *The Diamond Age,* and Houellebecq's *The Elementary Particles,* among other texts) claim that it promises "to destroy the meaning of our lives?"[24] As a form of anxiety with the power to overwhelm prospects of social change, this fear of technology is rooted in the forms of reflexive modernity that I have already begun to describe.

Reflexive modernity is only possible because of technology. The term "reflexive modernity" as described by Giddens characterizes the radicalization of the process of industrialization. As the principles of industrialization are analyzed by a society examining the consequences of its technological development, those principles become the object of reflexive critique. Beck's concept of the "risk society" relies on this definition of a technologically constructed reflexivity for the production of risk as a cultural force. Risks are tied to the probability of physical harm that results from human scientific and technological development. Thus, risk and reflexive modernity are mutually dependent: *"Risk may be defined as a systematic way of dealing with hazards and insecurities induced and introduced by modernization itself. Risks, as opposed to older dangers, are consequences which relate to the threatening force of modernization and to its globalization of doubt."*[25] For Beck, any critique of the science and technology that have produced these risks is an expression of reflexive modernity, which looks at the consequences of industrial society from within.[26] Novels that deal with nanotechnology, including Crichton's *Prey* and Stephenson's *The Diamond Age,* are critiquing its development through the production of nightmare scenarios. *The Diamond Age* profiles the threat of "microscopic invaders"[27] and describes "cookie cutter" nanoexplosives that detonate within the bloodstream of their victims.[28] Crichton's Jack Forman finds himself looking at "the tombstone of the human race."[29] Both are examining issues similar to those funded by the US government, when in August it provided $1m in funds from the National Nanotechnology Initiative for the study of the societal implications of nanotechnology, including its "unintended consequences." The RAND publication as well warns of "cultural threats" posed by nanotechnological development.[30] And *Jake 2.0*'s Jake Foley cannot walk through a metal detector without being physically disabled. Nanotechnology is clearly perceived as a risky business.

The reflexivity of nanotechnology as a risk technology extends to its economic environment. A recent National Science Foundation report on the "Societal Implications of Nanoscience and Nanotechnology" claimed that the worldwide market for nanotechnology-related products and services would top $1 trillion annually within 10-15 years.[31] The report cited business as the agent that will lead to nanotechnology achieving status as a "dominant force in society." Within the risk society, nanotechnology, as well as its accompanying critique, becomes a "system-immanent normal form of the revolutionizing of needs."[32] That is, the production of risk technologies produces a need for more advanced technologies to combat the self-produced risks. What better way to fight invasive nanomechanical devices than with a more advanced version? Or to put it another way, we can turn to still-imaginary blood-cleaning motorized devices to deal with plaque build-up in the aorta, arguably the by-product of industrialized society. Giddens' *Runaway World* requires risk for a functioning economy: "Risk always needs to be disciplined, but active risk-taking is a core element of a dynamic economy and an innovative society."[33] Risk must be produced, in effect, in the service of risk itself.

Don DeLillo's *Cosmopolis* posits exactly this problem. In a stretched limo, a venture capitalist and his advisor discuss the self-producing needs of technology-dependent cybercapitalism:

> "'It's cyber-capital that creates the future. What is the measurement called a nanosecond?'
>
> 'Ten to the minus ninth power.'
>
> 'This is what.'
>
> 'One billionth of a second,' he said.
>
> 'I understand none of this. But it tells me how rigorous we need to be in order to take accurate measure of the world around us.'"[34]

This rigor is doomed to inertia in the novel, in which self-reproducing technologies, beliefs, and institutions can be halted only by destruction. When the economic interests that fuel the initial research into a technological development then require its acceleration for sustainability in *Cosmopolis,* that system risks self-produced collapse.

The temporal operation that distinguishes nanotechnology as a risk technology is its future-orientation. The novels I have been discussing do not deal with the already destructive consequences of nanotechnological research; rather, they imagine its risk potential. The future component of risk initiates a cultural movement towards risk prevention. Speculative fiction holds free reign over possible near-futures for the technology. *Prey, The Elementary Particles,* and *Blood Music* do not create alternate worlds but instead relocate nanotechnology's future in the present. They utilize the future potential of the technology to examine the present, which becomes a product of its future rather than its past. The center of risk consciousness, says Beck, "lies not in the present, but in the future. In the risk society, the past loses the power to determine the present. Its place is taken by the future, thus, something non-existent, invented, fictive as the 'cause' of current experience and action."[35] Beck's causal interpretation of technology's future-orientation is problematized by any notion of a matrix-operation on the cultural formation of nanotechnology as a concept, but the fictional, future component of risk consciousness demonstrates the heightened status of wide-audience representations like *Prey* and *The Elementary Particles* (which was an international best-seller and multiple literary prize-winner) as projections of a future which, in effect, helps to create the technology.

The past-oriented temporality encoded by historical analysis becomes a target of reflexive modernity, which in Giddens' account only uses the past as a means of breaking with it. Instead, "'Futurology'–the charting of possible, likely, or available futures–becomes more important than charting out the past."[36] Instead of Walter Benjamin's angel of history being propelled backwards into the future, its face contorted in horror gazing upon the rubble piling up at its feet, the angel of the risk society is sucked, face-forwards, towards the future's irresistible gravitational pull. In *The Elementary Particles,* Houellebecq surprises the reader with the revelation in the final pages of the novel that the personal histories of two brothers that have been just been related were done so by products of a technological future. A new, "intelligent species," the result of Michel Djerzinski's multidisciplinary scientific approach to genetic engineering, calmly informs the reader that it (collectively) is offering the novel as a loving "tribute to mankind."[37] And while the novel's prologue includes the claim that the events in the novel lead to a paradigm shift, what that shift entails is not revealed until the brief epilogue. Houellebecq's framing device encircles the present of the novel, which gets radically redefined by its own future consequences. The technology-in-development of *The Elementary Particles,* a convergent relative of nanotechnology which does not bear the name, changes shape when faced with its own future.

Houellebecq does not explicitly ask for the future vision of his novel to be prevented, although the scientific development that forms the undercurrent of his narrative does have as its result the destruction of humanity as we know it. The question then arises whether a technology requires an implicit imperative to action to secure its status within a risk environment. In Niklas Luhmann's view, the concept of risk is an aspect of a decision formula, in which the unactualized future determines present decision-making processes: "Risk is therefore a form for present descriptions of the future under the viewpoint that one can decide, with regard to risks, on the one or other alternative."[38] This contradicts the movement of inevitability that functions in Houellebecq's description of a technologically enhanced human future. Luhmann's bestowal of agency on risk as a decision process is difficult to uphold when dealing with a technology produced by converging disciplines. Nanotechnology's potential and perception as a threat stems from its multiple causalities: so many people and institutions have their hands in it. Even Crichton's *Prey,* which addresses the need to consider the consequences of further developments in nanotechnology, wavers between a call to action and despair at the prospect of inevitability. When Jack Forman contemplates the processes which lead to the coupling of emergent behavior with technology, he worries about the inevitability of "self-optimization":

"But it hasn't been done with autonomous robots in the real world. As far as I know, this was the first time. Maybe it's already happened, and we just didn't hear about it. Anyway, I'm sure it'll happen again. Probably soon."[39] The fictional representations of nanotechnology have trouble descriminating between preventable and inevitable risk. As with the technology itself, the lines of agency are blurred, and risk as such emerges as the most powerful force in determining the nature of the nanotechnology represented in the popular forum.

The Architecture of Techno-Space

As a risk technology, nanotechnology reorients temporality, and in the process, it also realigns geographies. While it is the science of the very small, its scope is global. It has the potential to transcend the boundaries of bodies, buildings, and national borders, and those lines of spatial demarcation are intensely interrogated in the imagined futures of the technology. Furthermore, an aspect of risk is global reach—technological effects on the environment do not follow state lines, and individuals and their governing bodies must deal with anticipated risks that stem from technology they may or may not use. The territory of nanotechnological reach, as risk, is the entire planet.

Margaret Wertheim begins a discussion of nanotechnology for the Australian publication *The Age* by considering scalar possibilities: "As a realm of fantasy, outer space has always had the distinct advantage of size . . . Given enough space, one can imagine that somewhere almost everything must be happening." On the level of nanotechnology, what she calls the realm of "inner-space dreaming," the global reach of space begins at the smallest level and extends outward. The prospect of risk tied to the public concept of nanotechnology has both a temporal and spatial function. As a result, it is not surprising that architect Rem Koolhaas, in the June 2003 issue of *Wired* magazine that he guest-edited, included a section on "Nano Space" as one of the new spaces within what he described in his introduction as "the beginning of an inventory, a fragment of an image, a pixelated map of an emerging world."[40] That inventory also includes Bruno Latour's "Research Space," which looks beyond a decaying laboratory model of science towards a global lab which "has extended its walls to the entire planet."[41] The institutions of nanotechnology, global in reach, complement its perceived risk potential.

In 1964, Marshall McLuhan wrote that "after more than a century of electric technology, we have extended our central nervous system itself in a global embrace, abolishing both space and time as far as our planet is concerned."[42] Is the world of *Jake 2.0*, in which an "upgraded human" can plug into satellites and become a site of distributed intelligence, an abolishment or extension of space? It is impossible to consider nanotechnology as a phenomenon confined to US borders and pursue the same line of thought that leads David Nye, in his analysis of technology in *American culture in Narratives and Spaces*, to state that "Americans have appropriated and developed machines in their own way, and woven them into landscapes, social relations, and a sense of history."[43] This nationally specific sense of place and landscape cannot encompass the global reach of risk technologies. Lines of demarcation must be reconfigured.

When dealing with a technology small enough to render any fence obsolete, authors raise questions of permeability that lead some, like Stephenson, to return to the technology itself to deal with the issue of security. The defense system of *The Diamond Age* relies on a border patrol that mimics the body:

What worked in the body could work elsewhere, which is why the phyles had their own immune systems now. The impregnable-shield paradigm didn't work at the nano level; one needed to hack the mean free path. A well-defended clave was surrounded by an aerial buffer zone infested with immunocules—microscopic aerostats designed to seek and destroy invaders.[44]

This spatial immune system, which is effective but not impregnable, mimics the body's own boundary functions, which also come into question in the novel. Nanotechnology realigns perceptions of space as it redirects time towards the future. And those spaces are threatened spaces, spaces of potential harm from technological risk. From a geographical perspective, space flows from the scale of the body to the global. Edward W. Soja maps the infolding of border systems in a "postmetropolitan" world:

> Borders and boundaries are life's linear regulators, framing our thoughts and practices into territories of action that range in scale and scope from the intimate personal spaces surrounding our bodies through numerous regional worlds that enclose us in nested stages extending from the local to the global.[45]

Bodies, he says, are the centers of contextualized nodal regions. The properties of the nodal space of nanotechnology extend from the question of permeable bodily boundaries to permeability on the global scale. The concept of a spatial immune system applies as readily to a regional border as to a cellular one.

The global-local relay of nanotechnology finds its expression in the idea of miniaturization that led to nanotechnology's (inter)disciplinary conception. Richard Feynman introduced nanotechnology as a new field of physics by calling for the contents of the *Encyclopedia Britannica* to be written on the head of a pin. He also imagined a surgical scenario in which "You put the mechanical surgeon inside the blood vessel and it goes into the heart and 'looks' around . . . Other small machines might be permanently incorporated into the body to assist some inadequately functioning organ."[46] Feynman's idea of miniaturization finds its information-age proponent in Nicholas Negroponte who, at the beginning of *Being Digital,* says "the digital planet will look and feel like the head of a pin."[47] However, he later refines this idea by claiming that "the main reason for not putting something like a modem on the 'head of a pin' is no longer technological; it is that we have trouble keeping track of heads of pins and misplace them easily."[48] Space grows in the face of a miniaturized technology. When that technology becomes self-aware, as in *Blood Music,* it too has to arrive at a new conception of space: "They're trying to understand what space is. That's tough for them. They break distances down into concentrations of chemicals."[49] *In The Elementary Particles,* the creator of mechanized micro-biology not only "has clearly modified our perception of time; but his greatest contribution . . . is to have laid the foundations for a new philosophy of space."[50]

This "new philosophy of space" that accompanies fictional representations of nanotechnology begins with the body. The vision of nanoelectricomechanical systems (NEMS) inside the boundaries of the body is an interior mechanization, an appeal to agency very unlike the splitting of the atom. When Frederic Jameson describes the creation of "Spatial Utopias" in the 1960s, he says "the transformation of social relations and political institutions is projected onto the vision of place and landscape, including the human body."[51] The primary fictional landscape of nanotechnology is the human body, where the cellular transmission of DNA information becomes the model for the encyclopedia on the head of a pin. The bottom-up approach makes miniaturization a conceptual framework for under-standing nanotechnology, not the method for achieving construction at the molecular level. Michel Djerzinski of *The Elementary Particles* looks beyond what he calls "junk DNA" to find the simplest, and smallest, structures to work with: "There's nothing in principle to distinguish configurations prone to mutation, but there have to be some conditions for structural stability at a subatomic level. If we can work out a stable configuration with even a couple hundred atoms, it's just a matter of the power of the processor."[52] The construction that Djerzinski hopes to achieve (and does) atomizes the body at the nanoscale. It also turns the space of the body into a Koolhaasian "junkspace" of undif-ferentiated components that vary only in their arrangement. The architecture of the body becomes homologous to 20th century building:

> In previous building, materiality was based on a final state that could only be modi-fied at the expense of partial destruction. At the exact moment that our culture has abandoned repetition and regularity as repressive, building materials have become more and more modular, unitary, and standardized; substance now becomes pre-digitized... As the module becomes smaller and smaller, its status becomes that of a crypto-pixel. With enormous difficulty—budget, argument, negotiation, deformation—irregularity and uniqueness are constructed from identical elements. Instead of trying to wrest order from chaos, the picturesque is now wrested from the homogenized.[53]

The modular construction of the body, reduced in nanospace to the same materials from which mechanized devices are constructed in *Prey*, interrogates the difference between the two entities. Crichton's nightmare scenario is that the space of the two will cease to show any differentiation. His nanomachines learn to take part in the construction of the body, which Jack Forman witnesses as he magnetically separates the nanostructures from his wife: "The skin of her swollen face and body blew away from her in streams of particles, like sand blown off a sand dune . . . And when it was finished, what was left behind—what I still held in my arms—was a pale and cadaverous form."[54] In Crichton's vision, risk threatens the personal space of the body at the same time that it extends that threat worldwide. The elementary particles of all bodies, in this scenario, are equally subject to trans-formation at the nano level. The construction site of nanotechnology begins at the smallest level of the body and extends globally.

Nanotechnology, Fiction, and Sacrifice

Crichton's blurred boundaries between body and machine, however, do not blur the lines of identity. When the machines invade bodies in *Prey*, they either imprison the personality within it, or kill the person entirely by destroying the body and even breaking it down to use the atoms for self-repro-duction. By imprisoning the personality, the nanobots in *Prey* create a hybrid of body and machine, in which the original personality becomes dormant, and is represented by the machines in distorted form. That this destruction is self-imposed by human technology is also at issue. The question of nanotechnology for Crichton is to be considered defensively. As a risk technology, it initiates what Beck calls a kind of negative utopianism: "One is no longer concerned with attaining something 'good', but rather with preventing the worst; self-limitation is the goal which emerges . . . The utopia of the risk society is that everyone should be spared from poisoning."[55] But how does self-limitation function if the drive for progress and development within society, as he says, is not simply an ideology, but institutionalized?[56]

Crichton chooses self-limitation as the goal of his fictional representation of nanotechnology. While Bear's view is presented with more ambivalence, characters in *Blood Music* nevertheless tell each other to work hard to limit the development of the technology, "before we are all dead by our own hand."[57] *Prey* offers up progress in exchange for security, sacrificing technological development as an offering to reflexively produced risk. This policy of relinquishment takes up a negative position in the dialectical movement of nanotechnological development, and appears as a possible, but not necessarily advocated, protection of the material world's status quo in *Blood Music* and *The Diamond Age*. But something quite different happens in Houellebecq's *The Elementary Particles*. When Koolhaas, in "Junkspace," asks "Is each of us a mini-construction site? Is mankind the sum of three to five billion individual upgrades? Is it a repertoire of reconfiguration that facilitates the intromission of a new species into its self-made Junksphere?",[58] Houellebecq's overwhelming answer is yes, and why do anything about it?

The Elementary Particles, and to a certain extent *Blood Music*, go so far as to embrace the risk potential of convergent technologies. Houellebecq's novel envisions the paradigm shift that would result as total and transformative: "Once a metaphysical mutation has arisen, it tends to move inexorably toward its logical conclusion. Heedlessly, it sweeps away economic and political systems, aesthetic judgments and social hierarchies. No human agency can halt its progress."[59] This movement becomes a Bataillian form of expenditure, a sumptuous process of destruction with destruction itself as its goal. It is a game in which "the danger of death is not avoided; on the contrary, it is an object of a strong unconscious attraction."[60] Risk, for Houellebecq, runs counter to the principle of relinquishment. Instead, it works in the form of Bataille's potlatch, in which gain does not "serve to shelter its owner from need. On the contrary, it functionally remains—as does its possessor—at the mercy of a need for limitless loss, which exists endemically in a social group."[61] Houellebecq describes a situation in which humanity becomes "the first species in the universe to develop the conditions for its own replacement,"[62] in which technological development occurs in the explicit service of self-destruction.[63] This enormous undertaking—the demolition of the human world in the name of progress—is entered into willingly, almost deliriously. Houellebecq co-opts risk technology in order to sacrifice humanity in an extravagant fictional gesture.

The sacrificial drive that Houellebecq's novel embodies is described by Jean Baudrillard as a sacrifice to an experimental future; that is, the uncontrollable, limitless side of risk. Baudrillard dismisses the notion of any remaining instinct for self-preservation in the theater of scientific progress. The trend that has won, he says, is "the trend towards the sacrifice of the species and unlimited experimentation."[64] Baudrillard wants to describe something already, inexorably, in progress, while Houellebecq anticipates the consequences of converged technologies as risk technology in the present cultural moment. *The Elementary Particles* rushes headlong into a molecular, atomized,[65] sacrificial future that operationally reconfigures the present research community. The technology described becomes a deliriously positive risk, as opposed to a negative, self-limiting risk, yet it somehow doesn't seem very positive at all.

The constellation of meanings surrounding the social concept of nanotechnology threaten to sacrifice the concept itself to overdetermination, a destructive move that would not be unwelcome in *The Elementary Particles* or *Prey* alike. The future component of nanotechnology, governed by self-limitation for Crichton and limitless sacrifice for Houellebecq, in both cases gravitates towards destruction—either of the technology itself *(Prey)* or those that create it *(The Elementary Particles)*.[66] By eliminating both human progress and the human in one dialectical movement, these novels approach the limits of nanotechnological risk. At the point of destruction, Crichton calls for the sacrifice of developments in nanotechnology in order to save a pre-nano world. This implies that there is something inherent in the material conditions of the world as it exists before transformed by the technology that is, in effect, worth saving. It is at this point that Houellebecq's radical vision emerges as the braver, and in my view, more interesting of the two. The converged technologies in *The Elementary Particles* enact a purposive sacrifice of humanity and its material conditions, and this is done without any hint of regret. The problem of redemption that plagues Crichton's plague, the very fact that there is something which must be redeemed despite its folly, is simply not an issue for Houellebecq. What emerges instead is a technology which, in its selfish destruction of those that create it, reveals the nature of the creator. Why preserve a damaged humanity, he asks, when that damage is self-produced? By contemplating sacrifice on this scale, Houellebecq situates himself at the very scene of risk for nanotechnology: one where power, knowledge, and the problem of control converge. He does not back away, and that headlong rush is perhaps the most compelling way to grasp the future of this risk technology.

◇◇◇

Dust, Lust and Other Messages from the Quantum Wonderland
BRIAN ATTERBERY

In Greg Egan's story "Dust," later incorporated into his novel *Permutation City*,[1] a computer-generated character named Paul is subjected to a series of experiments in which his virtual environment is interrupted, run backward, and fragmented into a set of computations distributed among computers around the world. Ultimately he is so divided and randomized that he thinks of himself as having been ground into dust–and reassembled by his own perceptions. In each experiment, Paul's subjective impression is that both the world and his own consciousness are smoothly continuous, a continuity signaled by his counting to ten in each trial. Even though an outside observer sees Paul pausing or counting down from ten to one, for Paul each sequence is identical. Only he perceives the relativistic symmetry involved: "To an outside observer, these ten seconds had been ground up into ten thousand uncorrelated moments, and scattered throughout real time–and in model time, the outside world has suffered an equivalent fate."[2]

Egan suggests two possible implications for those of us who live in the non-virtual world. First, our own sense of an integrated self may also be an illusion–consciousness may be a way of covering up, smoothing over the gaps of which we are constructed. Second, the difference may not matter anyway because a sufficiently detailed simulation, whether of continuity, consciousness, or the universe itself, is as good as real.

Paul later discovers that he is not a computer program but a temporary immersion of the "real" Paul into a simulated sensorium. Returning to his own body, Paul asserts the legitimacy of the epiphany he experienced within the simulation. Further, he proceeds to theorize and then to invent a computational device that requires no hardware at all. Simply by running the first few seconds of a massive simulation on linked computers around the world and then suddenly cutting off the program, Paul proposes that he can generate a self-sustaining order as the conscious minds within the simulation continue to create order out of the random movements of the universe. "We perceive–we *inhabit*–one arrangement of the set of events," says Paul. However, he argues,

> There's no reason to believe that the pattern we've found is the only coherent way of ordering the dust. There must be billions of other universes coexisting with us, made of the very same stuff–just differently arranged. If *I* can perceive events thousands of kilometers and hundreds of seconds apart to be side by side and simultaneous, there could be worlds, and creatures, built up from what we'd think of as points in space-time scattered all over the galaxy, all over the universe. We're one possible solution to a giant cosmic anagram . . . but it would be ludicrous to believe that we're *the only one*.[3]

Paul's argument is nonsense at the level of ordinary experience, where we take for granted such basic data as sequence, continuity, and material existence. Much nonsense, though, becomes demonstrable sense at the submicroscopic scale. At the level of quantum phenomena, all sorts of impossible things start to occur. Objects are in two places at once, communicate without material connection, wink in and out of existence, and operate according to causes that follow their effects. The world described by quantum physics is, as George Gamow's fables demonstrated decades ago, a weird wonderland where common sense is just plain wrong. Indeed, Lewis Carroll's original wonderland, the model for Gamow's quantum parables, has become one of the most useful reference points for talking about the nanoscale world, as Susan E. Lewak points out in this volume.[4] A computer with no hardware, like the one Egan describes, is absurd on a human scale, but at the quantum level, it just might work.

Normally, we don't have to think much about the quantum wonderland. Whatever might be happening to the quarks and leptons of which our bodies are theoretically composed, all of that weirdness is canceled out by the time we reach the cellular level. The physicist Erwin Schrödinger achieved science fictional fame by inventing a thought experiment in which a quantum event—the decay of a single radioactive atom—might possibly reveal itself at something close to human scale. By bridging this conceptual gap, Schrödinger's cat, which remains neither alive nor dead until an observer opens the box and fixes the quantum indeterminacy—has become the unofficial mascot of contemporary hard science fiction. Variations of and embroideries upon Schrödinger's *gedankenexperiment* have been proposed by writers ranging from Rudy Rucker to Ursula K. LeGuin. The "Locus Index to Science Fiction" currently lists no less than thirteen different stories with titles giving Herr Schrödinger not only a cat but a kitten, a dog, a mousetrap, a fridge, and so on.[5]

Recently, though, Schrödinger's cat—and his other accessories—have gained some company. Developments in neurophysics, computer science, and nanotechnology provide other possible links between subatomic phenomena and the perceivable world. Nanomachines, quantum computers, and the components of the brain itself may transform everyday experience into a quirkier, quarkier realm. Egan is one of the explorers attempting to chart the new territory.

Up till now, most people could ignore the strange pronouncements coming from mathematicians and nuclear physicists. Of the two cultures described by C.P. Snow in 1959,[6] most of us who are not directly involved with scientific pursuits tend to function almost entirely within the limits of humanistic, non-scientific culture. This is especially true of literary critics, who stick to a standard humanistic discourse revolving around psychological dilemmas, social structures, and moral issues: all operating at the human scale and within the world of common sense. Within this discourse there is no way to speak of the submicroscopic world and its radically different principles. The one great exception among literary modes is science fiction. Like scientific popularizers, science fiction writers try to translate new formulations and discoveries into recognizable images and engaging characters. Unlike popular scientific writers, SF writers must dramatize, rather than simply explaining: discussions of writing within the genre often focus on getting rid of what are called "expository lumps" or "infodumps." Successful SF immerses the reader in a displaced world in which new ideas are implied within the characters' trials and tribulations.

As long as the scientific base for science fiction primarily concerned rocketry and astronomical observation—when the displaced world was Mercury or Mars—traditional plot structures and storytelling techniques were more than adequate to generate an emotional charge. Alien biologies and unearthly cultures could be added to the mix without requiring major innovations at the story level. *Robinson Crusoe* could be re-situated on Mars. Kipling's *Kim* could be translated into space opera. C.P. Snow's cultural divide still had to be bridged: the fiction reader had to learn the language and some of the methods of science, but the basic stuff of story carried over to the new cultural framework.

But what if science tells us we are already living on an alien world in which nothing is as it seems? What if the displacement involves not the characters' world but the reader's? Writers like Egan tell us, in effect, that we are Martians—that the world we think we know is in fact something quite different.

Egan's work often pushes the boundaries of traditional storytelling in order to challenge common-sense notions of the self and the perceivable world. In "Reasons to Be Cheerful" (1997), for example, the main character's damaged brain is repaired by microprocessors and specially tailored polymers, and as a result, he is first rendered immune to pleasure and then given a choice of which sensations will be enjoyable.[7] Will he like chocolate or cheese, prefer Bach or Beethoven, desire men or women? Happiness becomes completely elective, thanks to new technologies. In Egan's "Axiomatic" (1990), discussed by Brooks Landon in this volume,[8] it is ethics, rather than pleasure, that the character can plug in at will. In each case, basic human capacities are subject to nanotechnological fixes, and our understanding of who those characters might be is likewise disrupted. Our sense of the nature of *story* itself is undermined when central character can assume any sexuality, any pattern of behavior, any morality at will.

At the beginning of *Permutation City,* Egan's characters worry, in a more or less conventional way, about jobs and relationships, but as the novel progresses, they become more deeply concerned with creating artificial life, transforming themselves, and changing the scientific episteme. Many of them aspire to the condition of computer programs. They want to be liberated from death and the body, free to edit their own emotional states and memories. One of the main characters has a dying mother. The daughter's primary motivation is to get her mother uploaded into computer storage before her body goes.

Egan's novel takes us from a recognizable near-future earth to a barely imaginable virtual landscape inhabited by beings who are no longer really human. Though some of them think of themselves as the same selves we met at the beginning, they are now freed from bodily limitations, they possess godlike powers, and their primary temptation is a spiral into self-absorption followed by self-anni-hilation. Egan suggests that aspiring to program-hood can be a form of death wish. The dust that comes together to create an Edenic new world or a virtual Adam can also be the dust of dissolution and decay. Without the restrictions imposed by bodily existence, there is little difference between a conscious being and a random disturbance in the quantum matrix. Egan asks how and whether our human drives and our ability to connect with one another are accidents of scale, local phenomena like solid matter and sequence.

Geoff Ryman takes quantum storytelling in quite the opposite direction. Like Egan, Ryman has explored the implications of a disembodied existence in some of his fiction. The most striking example is the race of formerly human "Angels" who "slide" from star to star in the novella "A Fall of Angels."[9] The same Angels reappear incidentally in his novel *The Child Garden* (1989).[10] For Ryman, transformation into an energy being is not a matter of escaping physicality and its accompanying clouds of emotion. Being transformed into an Angel is, for the character Zee, a kind of amputation:

> Those miraculous eyes, those gliding joints, those delicate hands are no longer yours. No longer will you change the universe by simply grabbing it. Your family, your friends, the

whole mosaic of your memory belongs to someone, something, else. You hear no warm, close pumping, you are no longer reassured by your own subtle odors. You have no gender, or sense of smell or taste of touch.[11]

Only gradually does the condition become tolerable for Zee, and then only because another Angel teaches him to find joy in a new kind of physicality:

Bee showed me freedom and taught me to enjoy it, and for that I loved him.

We plunged through suns and swam through the canyons of the sea. Bee took me to a world made of crystal that was the size of a house. We fell like rain through the blazing depths of the substratum where there is no time and it is white, blinding white, from the explosion that is both the beginning and the end. Humans cannot go there.(3)

If Egan's characters long to be dust, Ryman's are driven by lust—a word with many shadings of meaning, from intense sexual focus to a wide-ranging appetite for, and joy in, life. Ryman's beings rarely spiral into solipsism, but instead are driven by lusts of various sorts to reach outward from the self. They desire to touch, to merge, to trade substance with other beings and the universe. The Angel Zee comes to feel pity for mere human beings, whose ability to touch stops at the skin: "Violence," he comments, "is the only way they have of breaking into each other besides sex"(29).

But that is only partly true. Science fiction takes us to strange new worlds, which turn out as often as not to show us something new and strange about the world we live in. When Ursula K. LeGuin created a planet full of hermaphroditic beings in *The Left Hand of Darkness*, she says that she was not predicting or demanding that humans should become androgynous; "I'm merely observing, in the peculiar, devious, and thought-experimental manner proper to science fiction, that if you look at us at certain odd times of the day in certain weathers, we already are"[12]

Just so, when Egan says that people could one day be made out of quantum dust, he is also showing us that we are dust. We come from it; we return to it; we force coherence upon it. When Ryman's Angels swoop through time and pass ecstatically through one another's bodies, he is saying something about sex and violence and the spiritual longing that is channeled through both. We do leak into one another imaginatively in fiction, emotionally in love, physically in the exchange of pheromones. Every individual is formed from the bodies of two others. We all continually slough off parts of ourselves into the environment. Our sense of an isolated, integrated self is partly a story we learn to tell ourselves, and yet storytelling is a code that invades us along with language in our first social interactions.

The belief that anyone can be isolated and self-contained is only sustainable within the narrow horizon of the human scale. Ryman's story suggests that, at certain times, in certain lights, we are Angels. Every joy that Zee learns to experience is, of necessity, expressed in language derived from earthly experiences and reflects back on those glimpses of transcendence. The quantum world leaks through into the human in the form of ecstatic union–or reunion–with the universe.

But ecstasy has a flip side. The lust for connectedness that drives Ryman's characters can lead to equally terrible agonies. The Angel Bee loses not only his lover but also his freedom. Having a physical body–even one made of light and motion–leaves one vulnerable. In Ryman's stories, scientific breakthroughs lead, as often as not, to new regimes of oppression, news ways to restrict and sunder and invade. His novella "The Unconquered Country," for instance, depicts new technologies of bio-engineering being used, not to free oppressed peoples, but to turn their very bodies into factories. The main character, a woman from a small, war-torn Asian country, is forced to use her womb as a nanotech assembler. "She was cheaper than the glass tanks," the narrator points out. The perfect no-overhead sweatshop, she can grow "parts of living machinery inside her–differentials for trucks, small household appliances."[13]

Here Ryman devastatingly transmutes Eric Drexler's vision of a nanotech paradise into an inferno by making it happen *in utero* instead of *in vitro*. Drexler's prophetic popularizations, such as the 1986 *Engines of Creation,* portrayed molecular assemblers remolding the world into an almost magical place of peace and plenty.[14] However, he did not show what effect his tiny machines might have on the human body, which is already an entire industry of such nanobots. Ryman insists on the presence of the body, with all its sensations. Or rather, not just "the body," but a specific body, the female, Asian, displaced, hungering body of an individual named Third Child. Third becomes, not just a factory, but a munitions factory:

> When Third was lucky, she got a contract for weapons. The pay was good because it was dangerous. The weapons would come gushing suddenly out of her with much loss of blood, usually in the middle of the night: an avalanche of glossy, freckled, dark brown guppies with black, soft eyes and bright rodent smiles full of teeth. No matter how ill or exhausted Third felt, she would shovel them, immediately, into buckets and tie down the lids. If she didn't do that, immediately, if she fell asleep, the guppies would eat her. Thrashing in their buckets as she carried them down the steps, the guppies would eat each other. She would have to hurry with them, shuffling as fast as she could under the weight, to the Neighbors. The Neighbors only paid her for the ones that were left alive. It was piecework.[15]

This passage works on many levels. As a metaphor for colonialism, it suggests how even a politically indifferent colonial (here, that of the invading Neighbors) preys on the very substance of subject peoples. As an exploration of gender politics it shows how women's bodies become territory to be claimed and exploited. As a study in survival, it demonstrates that living through war and oppression sometimes means a retreat to ever-shrinking parts of the self. Third Child successively loses her country, her village, her family, her lover, her personal freedom, and even control over her own body, but something remains inviolate even to her death and the story's end.

Reading the story at the quantum level, however, keeps us from finding too much affirmation in Third Child's struggle for integrity. The technology that allows Third to turn her womb into a munitions factory could be used to analyze and exploit other systems in turn. More layers might be stripped away: gene patterns patented; cultural heritage appropriated and commodified. The more deeply we delve into the patterns underlying human bodies and souls, the more easily the information can be transformed into exchangeable codes and thus controlled. Ryman explores some of these implications in *The Child Garden,* in which viruses, genetic codes, and computer programs become virtually interchangeable and only the hero's intractability saves her from absorption—the "bad grammar" that makes her virus-resistant, susceptible to cancers, attracted to other women, and an artist.[16]

Unlike earlier works such as "A Fall of Angels," *The Child Garden,* and *The Unconquered Country,* Ryman's 2001 novel *Lust* touches only incidentally on issues of knowledge and technology, and yet it can be read as his most explicit statement on the relationship between the quantum world and the realm of human experience. In *Lust,* the protagonist Michael experiences what he and the narrative term a "miracle."[17] He discovers he can conjure people into existence—can produce out of thin air duplicates of old lovers, attractive strangers, historical personages like Billie Holliday and Pablo Picasso. The only requirement is that he feel desire for the person on whom his duplicate is modeled. In the course of the novel, his London flat is occupied by a variety of desirable "angels," ranging from a thuggish guard on the Underground to a cartoon named Taffy Duck (a thinly disguised Jessica Rabbit, from the movie *Who Framed Roger Rabbit?).*

This, of course, is the premise of an erotic daydream rather than that of an investigation into the quantum basis of human thought. Michael manages at least the beginnings of an erotic romp, but he is not really the romping sort. His angelic encounters lead him gradually back, to the sources of his own emotional and spiritual malaise. Like one of the damned in Dante's *Inferno,* he starts devolving into a mere emblem of his own sins—especially the adolescent fixation on his own estranged father and the result self-condemnation that has kept him from forming a real attachment even to his long-time lover.

Much of *Lust* is thus not science fiction but fantasy—both sexual fantasy and spiritual quest. We almost forget, by the end of the book, that Michael is a scientist. But Ryman does not forget. When we first meet the protagonist, he is in charge of a laboratory where newly hatched chicks are dissected to determine the physical changes wrought by their first exposure to light Almost immediately after describing the procedure—killing, cutting brain tissue, staining, freezing samples—the narrative takes Michael outside for a lunch break, where he calls up on of the angels. He asks the angel, duplicate of a fellow he has previously nicknamed the Cherub, about the sources of the miracle.

> "It goes all the way back," the Cherub said. Then he turned and looked at Michael with a
> sudden urgency. "The back of the head." And he jerked it behind him.
>
> "You wouldn't happen to know what part of the anatomy?"
>
> "So far back it goes outside." (41)

Michael does not know what to make of this exchange, except to speculate that being summoned out of the air might cause mild damage to the brains of his angels. But the clues are there. His scholarly field is the intersection of neurology and philosophy: "the grey area where biology was helping philosophers answer questions such as: do we have a soul What is the self?"(18). Michael's chick experiment is an attempt to address the questions by examining their brains before and after a first exposure to light to see whether there is a physical structure to perception: a "grammar of sight" consisting of "verbs of movement, adjectives of colour, and nouns of space and shape"(18).

The eggs that are to hatch to provide test chicks arrive at the lab in a crate labeled "FRAGILE COMPUTERS" so that no animal rights protestors will be alerted(3). But the chicks are fragile computers, and so is Michael. Inside each skull is an amazingly sophisticated quantum computer, capable of transforming streams of photons into knowledge and action. As he tries to understand his private miracle, Michael recalls developments in quantum theory implying that objects could be completely read, and thus reliably re-created somewhere else. Or rather, duplicated. Michael had been searching for information on quantum computing and had accidentally ended deep inside the IBM website, on the page describing IBM's teleportation project. The aim was successfully to transport an inanimate object by 2050. There was the usual team of delighted, slightly skuzzy-looking men, thrilled to be living in the dreams of their youth.(33)

What Michael does not know yet, and so is unable to interpret the Cherub's statement, is that the capacity to see, to read an environment or a fellow being (at least one he dreams of), and to create duplicates, is already there in the brain. "It goes all the way back," as the angel said to him.

In the end, neither Michael's experiments nor his miracles tells him what he expects to hear. The chicks, it turns out, do form new neural pathways when exposed to light, but always the same pathways. The pattern is already there, implicit in the cells and the genes. One might say the lust for light is built in. In newborn chicks, as in the computerized art that the reborn Picasso has begun to produce, the code contains the entire image. The grammar already implies the sentence.

Michael's angels, too, already exist somewhere, waiting for his desire to invoke them. But where is that somewhere, and what is the process by which nothingness becomes substance? Michael realizes that there are precedents, and not just mythical ones. Space and matter come into being through natural processes, and there are entire fields of mathematics devoted to modeling those processes:

> Mathematics said there were eleven dimensions in all. Four of them existed only in time. The three dimensions of space were created by the big bang. They were expanding outwards. That expansion was simply time, flowing in one direction only: towards the future.

> But.

> That would mean there were seven dimensions outside time. They would be just as small as they were before the big bang. They would be a point. No height, width or depth. They would be like the smallest dot made by the sharpest pencil. But that dot would be everywhere. It would be at the core of everything around you. It would be in the core of you. You live there, but don't know it. Everything in your life flows in one direction only, into it.

> A word came to him: neurophysics. The extension of the self into the universe. (375)

Michael's thinking at this point is full of analogies and metaphors: the self (or soul) is like a white hole; the flow of time is like light passing along the optic nerve; desire, like gravity, pulls things into being. However, the understanding he reaches cannot be reduced to metaphor, as if the entire experience were an extended parable designed to teach us that love is a force or imagination a way of making something out of nothing. If that were the case, all the talk about neural nets and non-spatial dimensions would be nothing more than television sci-fi technobabble.

In science fiction, as in science, the choice of model matters. Once chosen, the logic of the model must drive the story. Michael's neurophysics is based in real phenomena and real scientific attempts to understand them. This is not to say that Ryman is performing real *gedankenexperiments*. There are a lot of holes (and not just black and white ones) in Michael's hypothesis. Though mathematical models are invoked, no math is actually being done in the story. Science fiction functions somewhere between the literal and the metaphoric. It depends on unresolved suspension between the two.

Greg Egan, too, leaves holes in his fictional science, although he camouflages his more elaborately. In the initial "dust" simulation, for instance, there is no way for the computers to simulate Paul's count of "three" without information about the prior states of "two" and "one."[18] Yet the impossible premise permits Egan to explore real scientific questions and their impact on humans (or their virtual successors). Even if Egan's no-hardware computers are impossible, they are still elegant fictional analogues for the mysterious processes by which randomness generates order and matter becomes self-aware.

Ryman's balancing act between humanistic metaphor and scientific literalism likewise allows him to generate stories that interrogate the mysteries of desire and creation. Furthermore, these fictions find new ways to represent those mysteries as upwellings from the quantum world into the "real" world of perception and human interaction. "It goes all the way back," as the angel says, to the improbable properties of light and gravity.

Both writers are finding ways to harvest new and powerful stories from the seemingly remote speculations of mathematicians and cosmologists. In place of Schrödinger's cat they offer us airy computers, modularized brains, and angelic passions as new vectors for messages from the quantum wonderland.

Ryman and Egan are by no means the only writers exploring the quantum underpinnings of human existence. Probably the richest fictional exploration to date of the transformative potential of nano-technology is Kathleen Ann Goonan's four-volume sequence beginning with *Queen City Jazz* and ending with *Light Music*. Goonan offers an array of possible futures combined into one, with re-engineered humans, hive minds, artificially grown islands, and people who are living radio receivers. But the most striking single image is a vision of "enlivened" cities, their buildings topped by gigantic flowers and swarming with human-sized bees. Bees and flowers together constitute a system for gathering and exchanging information. The bees collect metapheromonal pollen, and the information encoded on the pollen is stories:

> Verity looked up and saw a few Bees scattered here and there in the sky. Ah, how information had become mixed. What were *they* doing with it, those alien creatures whose eyes polarized sunlight; why were they in charge of moving it from building to building, and how did they do it? . . . Why *were* they packing stories into the pollen baskets on their legs and carrying them to and fro? Had they absorbed so much of what being human was that they craved it?[19]

What the bees crave is what we all crave: the order and significance that only stories can give to the facts of existence. Those facts imply that the substructure of the universe is random noise: dust. Greg Egan shows how dust can rise above itself and seek to know itself. Other facts suggest that space, time, matter, and energy all were pulled into existence by hidden forces that still operate within our bodies and brains. The only way we know those facts is through those same bodies and brains. When we use metaphors—drawn from embodied experience—to help us understand one realm, we may find that they help us understand the other. As Ryman's protagonist realizes, "I can account for the yearning between stars. Somewhere in all that process is yearning between people as well"[20]

Stories of lust and stories of dust: both convey messages from the quantum world, which is also the world in which we have been living all along without knowing it. We crave those messages because they fertilize the imagination. Pollen is, after all, a powder that encodes the reproductive impulse of plants: it is both lust and dust. Flowers have been sending nano-scale messages to one another for eons; only recently have humans learned to intercept those messages and to read the stories encrypted therein. Like Goonan's bees, we are addicted to story-bearing pollen because we need to know who we are and what we might become. Stories like Goonan's, Ryman's, and Egan's offer new answers to some very old questions.

○ ○ ○

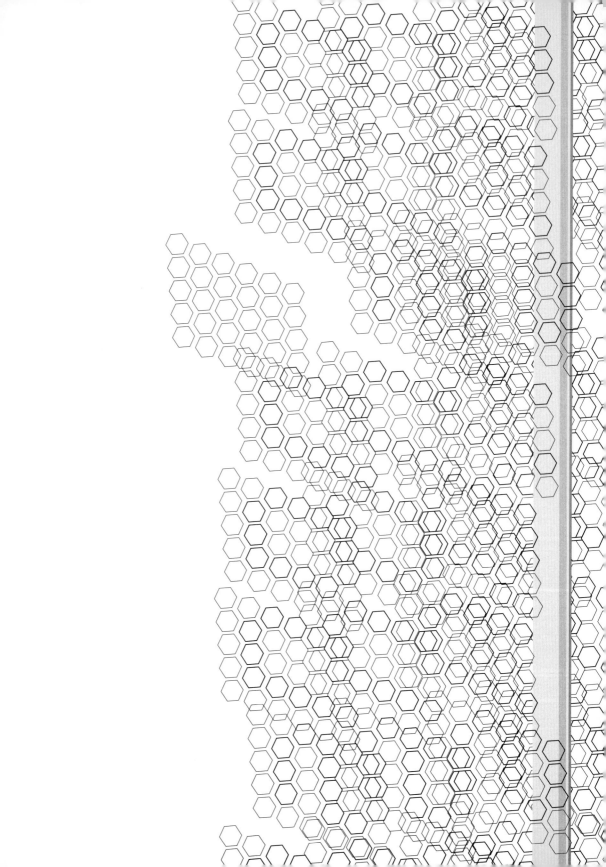

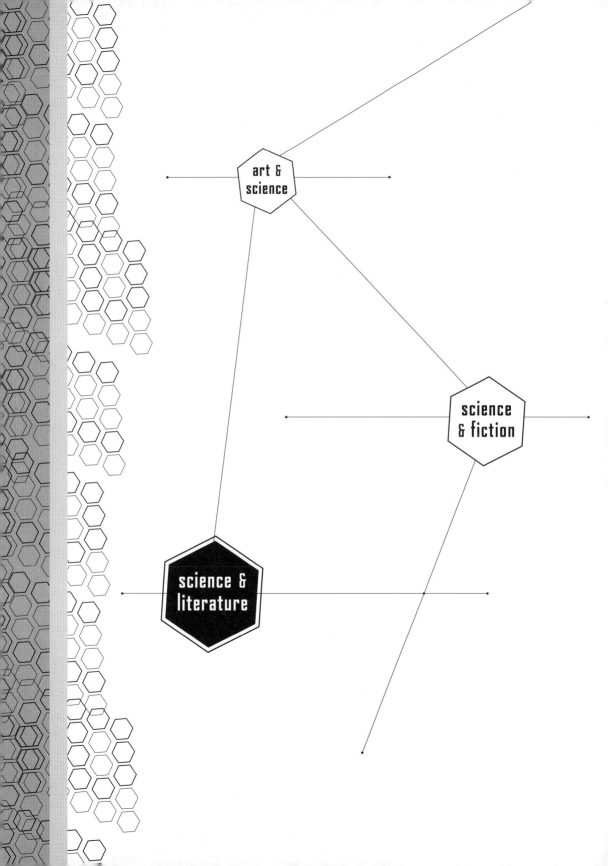

art &
science

science
& fiction

science &
literature

Needle on the Real:
Technoscience and Poetry at The Limits of Fabrication
NATHAN BROWN

Interface

In his 1959 address to the American Physical Society, "There's Plenty of Room at the Bottom," Richard Feynman posed "the final question" of "whether, ultimately—in the great future—we can arrange the atoms the way we want; the very atoms, all the way down!"[1] When Donald Eigler and Erhard Schweizer of IBM Almaden Labs answered Feynman's question thirty years later, by discovering a method for positioning single atoms using a scanning tunneling microscope (STM), one of the first tasks at which they tried their new technique was writing. Depositing thirty-five xenon atoms onto a substrate of nickel in an ultra-high vacuum cooled to liquid helium temperature (269 degrees centigrade below the freezing point of water), Eigler and Schweizer manipulated the forces of attraction existing between the nanoscale tip of the STM and the individual atoms with which it was brought into proximity to slide the atoms along the surface, one-by-one, until they had spelled out the letters I – B – M with precision control.

Courtesy of IBM: http://www.almaden.ibm.com/vis/stm/images/stm10.jpg

Noting that "it should be possible to assemble or modify certain molecules in this way," Eigler and Schweizer declared in their 1990 report to Nature that "the possibilities for perhaps the ultimate in device miniaturization are evident."[2]

The project had a long foreground. The immediate challenge issued in Feynman's inaugural talk was to render the information on the page of a book 1/25,000 smaller in linear scale, "in such a manner that it can be read by an electron microscope."[3] That challenge was met in 1985, when Stanford graduate student Tom Newman programmed an electron beam apparatus to inscribe the first page of *A Tale of Two Cities* in the appropriate dimensions.[4] Each of the letters in Newman's text, however, was approximately fifty atoms across, and Feynman's speculations on the possibility of atomic positioning presupposed his call in the same paper for better electron microscopes that would enable one to see individual atoms in the first place. And although transmission electron and field-ion microscopes have enabled atomic imaging under restricted conditions, the limitations imposed on electron microscopy at the atomic scale by electron diffraction, lens accuracy, and the damage inflicted by electron guns upon the sample itself eventually necessitated an altogether different approach.[5] Invented in the early 1980s by Gerd Binning and Heinrich Roher, who received the 1986 Nobel Prize for their work, the scanning tunneling microscope solved the optical problems of atomic imaging by having done with wave-optics altogether, in favor of a "tactile" interface. As Eigler puts it in a recent paper, "incongruously, [the STM] forms an image in a way which is similar to the way a blind person can form a mental image of an object by feeling the object."[6]

The basis for the operation of the STM is the quantum mechanical phenomenon of electron tunneling that occurs when a conducting needle, narrowing to a single atom at its tip, is brought into close proximity with a conducting or semi-conducting surface. A current is established at this interface by applying voltage between the tip and sample, and since the magnitude of that current is minutely sensitive to the distance between the two conductors, it can be used to establish a feedback loop that will adjust the position of the tip in accordance with the topography of the sample. Mounted on a piezoelectric transducer that adjusts its height with finite control, the tip is scanned across the surface, rising or falling according to the atomic terrain it encounters. From the information gathered by this "tactile" sensitivity, a heavily mediated visual map of the sample's atomic structure is digitally constructed and displayed on a monitor.

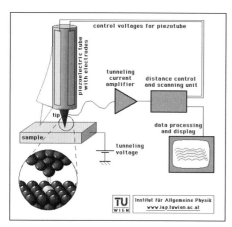

In order to manipulate individual atoms that have been deposited on a substrate, one has only to lower the tip to the point at which the Van der Waals and electrostatic forces existing between *tip* and atom are sufficient to overcome those between atom and *surface,* but not so great as to "pick up" the atom entirely. Within this range, the atom can be "dragged" across the substrate, and the change in location can be recorded by returning the tip to its initial height and re-imaging the sample.[7]

Courtesy of Institut für Allgemeine Physik: http://www.iap.tuwien.ac.at/www/surface/STM_Gallery/stm_schematic.html

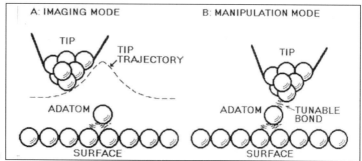

Courtesy of Eigler (1999)

Since Eigler and Schweizer's breakthrough inscription in 1989 and the legion of copycat efforts that followed, the production of so-called nanoscale "graffiti" through proximal probe lithography has taken a backseat to the engineering of atomic switches, molecular propellers and abacuses, and "quantum corrals"–atomic enclosures designed to confine electrons within a space where their wave properties can be studied.[8] But whether or not the STM has been programmed to spell out the word "PEACE" or to herd single atoms into an anagogic advertisement for a universe of nothing but International Business Machines, it always writes. As an inscription technology (a device that "initiate[s] material changes that can be read as marks"[9]), the STM constitutes an event in the history of writing machines insofar as it makes marks on a scale *beyond* optics, at which visual (re)presentations are predicated upon the radical priority of a haptic interface. And as Lisa Gitelman points out in *Scripts, Grooves, and Writing Machines*, "new inscriptions signal new subjectivities."[10] What new subjectivities are latent when technoscience arrives at what those in the business call "the limits of fabrication?" Rewinding Kittler's genealogy of gramophone, film, typewriter–with a difference–the STM returns us to the stylus as the locus of inscription, dropping a needle on the real in a realm that is not only "humanly," but *technically* invisible.

Face to Face

Poetic approaches to the limits of fabrication are not so historically determined. Sometime around 1862, Emily Dickinson starts a poem with "I cannot live with You –", then proceeds to unfold a labyrinth of grammatical, theological, and syllogistic implications before arriving at the following decisive formulation: "So We must meet apart – / You there – I – here –".[11] Writing can go no further.

"You there – I – here –": the first thing to notice about this line is that, along with Dickinson's trademark dashes, it is composed entirely of deictic terms, or "shifters." The dash is a minimal graphemic unit–pen touching down on paper in an instant's motion, leaving the barest trace of furtive contact. Shifters are the piezoelectric transducers of grammar–minutely sensitive to the voltage of voice, expanding to generate an illusory fusion of "body":"language":"world" at the interface of the tongue's tip.

Shifters include pronouns ("You" and "I" in this case) and also indices of spatio-temporal location–"there" and "here" in Dickinson's line, but also "now" and "then" or "this" and "that." Roman Jakobsen explains that shifters "are distinguished from all other constituents of the linguistic code solely by their compulsory reference to the given message."[12] In other words, they have the indexical function of establishing an existential relationship between a particular subject and object, place and time, and a particular speech act. As Emile Benveniste puts it, shifters constitute "an ensemble of 'empty' signs that are nonreferential with respect to 'reality.' These signs are always available and become 'full' as soon as the speaker introduces them into each instance of his discourse."[13] Giving the concept of the shifter something of a philosophical workout in his famous chapter on "sense-certainty" in the *Phenomenology of Spirit*, Hegel addresses the implications of these "empty" signs for written language:

> To the question: "What is now?", let us answer, e.g. "Now is Night." In order to test the
> truth of this sense-certainty a simple experiment will suffice. We write down this truth;

a truth cannot lose anything by being written down, any more than it can lose anything through our preserving it. If *now, this noon,* we look again at the written truth we shall have to say that it has become stale.[14]

"Here" and "now," the spoken shifter anchors a subject in a context. Later, re-activated by an act of reading, the context of the written shifter shifts. Some one hundred and seventy years after Hegel's meditation, Roland Barthes would seize upon precisely this slippage between spoken and written shifters as "a precious analytic instrument" for his "destruction of the Author."[15]

If, in the course of reading Dickinson's poem, one relinquishes the impulse to ask after the identity of the "You" with whom "she" cannot live, that's because it becomes increasingly clear that the poem might as well be "about" the status of pronominal reference itself. Stanzas eleven and twelve, for example, seem at least as concerned with the interpersonal dynamics of deixis as they are with the vagaries of the afterlife:

> And were You lost, I would be —
> Though My Name
> Rang loudest
> On the Heavenly fame
>
> And were You — saved —
> And I — condemned to be
> Where You were not —
> That self — were Hell to Me —

In the first line, "I" is exposed as an utterly meaningless category without some "You" to whom it can be spoken, and the second line confirms the hollowness of the proper name in the absence of the personal pronouns that link us with language. Benveniste points out that "consciousness of self is only possible if it is experienced by contrast. I use *I* only when I am speaking to someone who will be *you* in my address...It is this condition of dialogue that is constitutive of *person.*" And again, on the next page:

> Language is possible only because each speaker sets himself up as a subject referring to himself as I in his discourse. Because of this, I posits another person, the one who, being as he is completely exterior to "me," becomes my echo to whom I say you and who says you to me. This polarity of persons is the fundamental condition of language.[16]

It's no wonder, then, that if "I" were "condemned to be / Where you were not — / That self — were Hell to Me —". Dickinson's "solution"–meeting apart–preserves the exteriority of "You" while saving "Me" from a solipsism bereft even of language. In his 1976 collection–*Shifters*–Steve McCaffery sums up the cut that necessarily connects like so:

> you
> are what
> i
>
> am apart
> from

what

i

is

a part

of [17]

"You there – I – here –": without reference to Dickinson, McCaffery hones in on the undercurrents of her line in a note to his volume: "shifters shift within a topography and topology of text where every 'i' is a 'here' and every 'you' a 'there'. poems then of openness and closure. semiotic bars and semiotic centres unfolding as tests of their own meanings."[18] "Semiotic bars" returns us to Dickinson's dashes, opening a passage to the second feature of her line that I want to highlight. The line takes on its full significance only when we recognize it as a paragram. Leon Rodiez explains that a text is paragrammatic "in the sense that its organization of words (and their denotations), grammar, and syntax is challenged by the infinite possibilities provided by letters or phonemes combining to form networks of significance not accessible through conventional reading habits."[19] In Dickinson's line, the paragram operates on an even smaller scale: at the level of the grapheme. The second half of the line, "– I – here –" emerges from the graphemic elements of "there." In print, the "t" splits both vertically and horizontally, transforming "t" into "I" and also forming the dashes that both separate and conjoin "I" with "there"–and with its remainder, "here." But this paragram is even more evident in the fascicle edition of Dickinson's poem, in which the "t" is not crossed, but rather looks precisely like the vertical stroke of "I" with a horizontal stroke just to the right of it:

[20]

The "crossbar," that is to say, is already a dash which has literally been placed *over* "there," before, between, and after which it disperses itself to reveal the latency of "– I – here –" within that word.[21]

Dickinson's line begins to resonate ontologically at the intersection of the two constitutive properties that I've mentioned: the paragram enacts a sifting of shifters in which our primary existential place-holders are made manifest and held in proximity by the rupture of a grapheme. What work do we have cut out for us as readers of such a line? In what *dimension* does a paragram "occur?" "Para-" means beside, alongside of, past, or beyond. The root form, *per-*, connotes forward, through, in front of, first, toward, against, near–*to pass over*. This network of associations constitutes the locus of Emmanuel Levinas's thinking in *Totality and Infinity* and *Otherwise than Being,* in which the proximate exteriority of the You *(autrui)* to the "I" is named the "face to face." For Levinas, ethics precedes ontology insofar as the relation to the Other at once calls the I into question and into being. What Levinas calls "the idea of Infinity" is this formal structure of the face to face, in which the "strangeness of the Other"[22] evokes the approach that is language and the overflow that is responsibility. As I've been reading it, "You there – I – here –" is a striking formalization of the face to face.[23] But while Levinas's thought is permeated by a certain anthropocentrism ("things have no face" [140]) and a strong phonocentrism ("speech is thus the origin of all signification" [98]), Dickinson's I and You seem to emerge not only from the encounter of self and Other, but also from an unstable intersection of pen and paper, body and technology, "language" and writing.

Perhaps one could read Levinas against Levinas by expanding the "idea of Infinity" to include such configurations,[24] in order to argue that an ethics of the approach would involve a responsibility to not only to the "You" but also to the "It." The STM writes at the interface towards which "Dickinson" gestures, where the visible "topography and topology" of a text emerge from an overdetermined graphemic intervention. We might imagine the "scene" of its writing as the very non-space in which the visible *breaks down*—that unthinkable site "between" the "there" and "here" of Dickinson's line. At the limits of fabrication, poetry and technoscience operate at the horizon of the visible and beyond, in those *impossible, illegible* spaces from which we are approached by bodies and words, and wherein text passes over into texture.

Fabrication

Fabrication designates the field in which technoscience and poetry come together under the sign of building, as branches of *materials research*. Poet Robert Smithson argues that "from the linguistic point of view, one establishes rules of structure based on a change in the semantics of building."[25] For Smithson, the "semantics" of building have a tendency to crack, opening into sediment. "Look at any word long enough," he invites us, "and you will see it open into a series of faults, into a terrain of particles each containing its own void." Taking up the investigation of such minute poetic particulars, along with the emphasis on the materiality of the signifier that continues to redefine literary studies, avant-garde critic Craig Dworkin calls for "radical formalisms" that "hew to the concrete. Where 'concrete' is what the street is made of."[26] But although concrete—that tried-and-true mixture of gravel, sand, cement, and water—exemplifies a *synthetic* mode of macro-building to which nanotechnology is distantly related, materials research and fabrication at the nanoscale is dedicated to synthetics of a slightly different order. "The milestone in man's ability to build things" heralded by the achievement of atomic positioning "is the ability to build things using individual atoms as the building blocks; the ability to build things *from the bottom up*, by placing the atoms where we want them." Against the relatively ad hoc structure of humble concrete, "atomic scale construction embodies the idea that the structure is exact in the sense that (within manufacturing tolerances) we build just exactly what we want and nothing else."[27]

To fabricate is to "make something up": to construct or manufacture; to frame or invent. To forge. The verb comes from the Latin noun *fabrica,* or fabric, which involves the *textural* implications we've already encountered. A fabric is "a woven stuff," a "contrivance; an engine or appliance," "a product of skilled workmanship," or "any body formed by the conjunction of dissimilar parts," especially with reference to the animal body.[28] The Indo-European root, *dhabh-*, means "to fit together," yielding "daft" (from the Old English gedæfte—mild, gentle) and the Germanic *dab-:* "to be fitting" or "becoming." So to construct or concoct, contrive or invent—to engage in artifice—is both daft and only fitting. All in all, it becomes us.

As we know, poetry is a mode of making. In the Heideggerian version of this rumination, *poiesis* is construed as that kind of *building* that enables us to dwell, and since dwelling is characterized by its allegiance to sparing and preserving, it puts fabrication under the condition of ethics. When Heidegger asks the question concerning technology, he finds that *techne* and *poeisis* are both modes of *aletheuein,* or revealing, but insofar as modern technology has imposed itself in the mode of enframing—setting upon nature as standing-reserve—it has compromised the mutual implication of building and dwelling that Heidegger uncovers in the etymology of *Bauen.* Poetics and technics implicate each other, but *poeisis* (as a mode of bringing-forth rather than challenging-forth) takes precedence over *techne* to the extent that we aspire not only to fabricate, but to "dwell poetically." Hot on the trail of Hölderlin and proximate to Smithson, Heidegger declares that "authentic building occurs so far as there are poets, such poets as take the measure for architecture, the structure of dwelling." Initially a "kind" of building, poetry is finally designated as "the *primal* form of building."[29]

For media and technology theorist Mark Hansen, such claims make Heidegger guilty of the sin of *technesis:* "the putting-into-discourse of technology" by which technology's "robust materiality" is supposedly assimilated to "thought" and to language.[30] Doggedly tracking manifestations of *technesis* from Heidegger through varieties of what he somehow manages to call "poststructuralist representationalist ontology" (ET, 8), Hansen calls for critical models that will value the "primacy of embodiment over cultural construction" (ET, 51), attending in particular to the "unmarked alterations" that technologies "operate on our basic perceptual and sub-perceptual experiential faculties" (ET, 2). Heidegger, he argues, subordinates technology to the world-disclosing power of *Dasein* (ET 117) thus reducing it "to the status of pure instrumentality" and ultimately discounting its threat to Being itself (ET, 121).

While I concur with Hansen's emphasis upon embodiment and with his argument for the necessity of acknowledging technology's exteriority to the human, I am largely unsympathetic to his critique. Beyond his frequently willful misrepresentations of the thinkers under investigation,[31] the problem with his analysis is that, by consistently opposing "robust materiality" to language and to inscription, he paints himself into the corner of having to associate all engagements with the materiality *of* language as a disguised incursion upon the exteriority of technology—indeed, as a disguised humanist logocentrism. Thus, he describes embodied experience as "noncognitive,[32]" as though cognition were not embodied, and he argues as though an attention to "the real operation of machines"[33] is somehow compromised by the recognition that language (and particularly writing) *is* a technology that is also technologically mediated, and which constructs us even as it enables our constructions.

It's hard to say what "real machines" Hansen has in mind, but when we consider the STM, inscription is no humanist metaphor, nor does an engagement with "inscription technologies" delimit the agency of these (or technology in general) to a range of purposive "applications." Through the monitor upon which its data is displayed, the STM writes "for us." At the interface between tip and sample, it marks the real in a manner that is radically inaccessible to the human sensorium. Insofar as the "haptic" order of that interface has fundamentally altered microscopy as such, it alters not only the future of our "perceptual and sub-perceptual faculties" but also the future of the technological real itself—the configuration of those machines that will come into and occupy the world in their "robust materiality."

At the limits of fabrication, what is worth retaining from Heidegger's questioning of technology is his putting-into-technology of discourse—his sense of the extent to which *techne* includes *poeisis*—and his insistence upon the exteriority of *both* to the dimension of the human and of thought: "the essential unfolding of technology gives man entry into something which, of himself, he can neither invent nor in any way make;"[34] "man acts as though he were the shaper and master of language, while in fact language remains the master of man."[35] What we should *not* retain—as Hansen rightly argues—is the split that Heidegger opens between *modern* technology and poetry, and the redemptive story of the "saving power" that can be drawn *from* the danger of modern technology (as setting-upon) *by* poetry (as bringing-forth). To be sure, nanotechnology operates at the limits of enframing, promising "nearly complete control of the structure of matter."[36] But so does what Charles Bernstein calls official verse culture, setting upon language as a standing-reserve at the beck and call of an expressive subject. Heidegger's formula for the exterior agency of language is "language speaks," but what does the "talking" when Emily Dickinson writes "You there – I – here –"? Language : grammé : body : technology : space : the Other speak. There is no way to decide in such a case, and "poetry" cannot be so distinguished as to constitute the "primal" form of fabrication.

Approaching this zone of indiscernibility *from* the field of poetics, Steve McCaffery designates it as the domain of the "protosemantic":

> the protosemantic is more a process than a material thing; a multiplicity of forces which, when brought to bear on texts (or released in them), unleash a combinatory fecundity that includes those semantic jumps that manifest within letter shifts and verbal recombinations, and the presyntactic violations determining a word's position: rupture, reiteration, displacement, reterritorialization. It is also the invisible in writing, that which looks at us without actually appearing itself. Like the paragram, it remains invisible but is already there, establishing an uncanny position from which we are scrutinized by language.[37]

Again, I would stress the instability even of "language" within such a quantum-mechanical borderland, and McCaffery himself is hardly reluctant to do so.[38] Defining writing as "a material scene of forces,"[39] and reaffirming the Lucretian equation of letter and atom in order to urge "a serious consideration of both a residual and a possible micropoesis,"[40] McCaffery puts "poetics" under the condition of "tech-

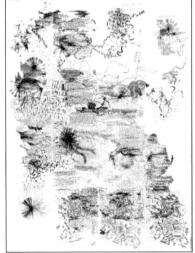

noscience" by making the historical materialist argument that "in our age of incipient miniaturization, it might be apt to return to the rumble beneath the word."[41] The word "nanopoeisis" will not be appearing outside of scare-quotes in this paper, but we can nonetheless observe that a protosemantic approach to making becomes even more apt when our "age" moves from the top-down fabrication procedures of "micro-" to the bottom-up methods of "nano-."

McCaffery's own technopoetic practice is perhaps the best instantiation of the theory that he preaches. With its mail-stamp mandalas, figure fives, painstakingly stenciled signifying chains, paratactic collisions of textual fragments, illegible overwritings, and dendritic-lettristic-galactic interzones, The *Second Panel* of McCaffery's *Carnival* offers a stunning vista over the protosemantic fabric:

42

"Vista over the protosemantic fabric" is meant to evoke the borderland between visuality and tactility, text and texture into which such a work draws us. As McCaffery explains in a 1998 interview, "*Carnival* is closer to cartography, to a diagram or topological surface than a poem or 'text'." "The panel when 'seen'," he notes, "is 'all language at a distance'; the panel when read is entered, and offers the reader the experience of non-narrative language."[43]

In *Reading the Illegible*, Craig Dworkin argues that "to speak of the 'purely visual' nature of such texts...or to close off their *reading* by classifying them as 'visual' art, would be a mistake—and not just because writing is itself (always) already a visual art."[44] What Dworkin gestures toward here is the tactile dimension from which the visual emerges. To enter into such a dimension, where seeing, reading, and writing collude, is to practice what Jed Rasula calls "wreading": "the mulch-work in the nutritive ecology of the ground in which we attain a migratory prowess by detaching the w from writing and attach it, prosthetically, to wreading." "Wreading is the specular prosthesis of the text."[45] If the STM was said to write somewhere within the impossible domain of Dickinson's paragram, we can wread *Carnival* somewhere *into* the protosemantic fabric of the STM's writing—into that zone in which "we touch, therefore we see."[46]

But although McCaffery's piece was constructed between 1970 and 1975, this is no demonstration that *poeisis* is prior to, or more "primal" than, *techne* as a mode of building. The network of technical mediations upon which *Carnival* is predicated precludes such a conclusion,[47] and McCaffery's reference to "micro" rather than "nano" poeisis suggests that the conceptual tools proper to wreading science and literature into each other are subject to a *bi-directional* technocultural lag. If poetry, occasionally, opens those imagined spaces that "knowledge never knew" (to cite another McCaffery title), then technoscience, eventually, provides us with a new model of fabrication by which to know what we imagined. And such new knowledge spurs fresh imaginations.

It's in the midst of this onto-epistemological flux between science and poetry that I want to place the work of two contemporary poets working at the brink of "nanoculture." For Christian Bök and Caroline Bergvall, neither technoscience nor poetry is going to "save" anything, but each impinges upon the other as an inescapable latency that may or may not be brought to the foreground of fabrication.

(sur)rational solids, (sur)realist liquids

I should be explicit about the nexus of formal and readerly concerns that I think our engagement with nanoscience and technology might bring to the foreground of contemporary poetics. Thus far, the "haptic," quantum-mechanical interface by which the scanning tunneling microscope produces quasi-objects for scientific interrogation has served as a figure for the invisible paragrammatic effects and protosemantic textures that link language to bodies and technology within the field of what is normally called poetry.[48] And I have wanted the STM to register not *only* as a figure, but also as the putative harbinger of a broad and consequent shift that nanotechnology may or may not operate upon the economy of sensory modalities in Western culture (emphasizing—or manufacturing—the priority of tactility to visuality). In this last respect one might say that I have been concerned with what Merleau-Ponty calls "the touch-vision system"[49] as it operates in nanoscience and poetry. So far, my poetic examples have been deliberately anachronistic—"residual" rather than "possible," in McCaffery's terms. While shifts in wreaderly attentions are substantially historically determined, they do not respect linear history in their migratory habits. Approaching the limits of fabrication demands both the greatest precision and a high tolerance for the profound disorder of quantum effects—an ability not only to toggle between the pristine structure of "You there – I – here –" and the chaotic dissemination of *Carnival*, but to inhabit that dimension in which they come together. When Dworkin writes of a radical formalism, such conjunctions are what he has in mind, and the attention to minute particulars that writing and reading poetry have always demanded is radicalized when we engage those activities as modes of condensed matter research, wherein the fabrication of biotechnical linguistic systems is practiced and studied from the bottom up. To do so is not to concede "the surrender of culture to technology," as Neil Postman's unfortunate subtitle has it.[50] Nor, on the other hand, does it require reference to nanotechnology at all. I am simply arguing that this intersection of nanotechnology and the protosemantic constitutes the zone in which science and poetry impinge upon each other "today."

As nanotechnology begins to percolate out of the laboratory and into the lifeworld in the upcoming decades, the conjunction of order and chaos that fabrication at the nanoscale necessarily involves may be most evident in the merger of physics, solid-state chemistry, and molecular genetics. In Charles Lieber's estimation, the "grand challenge" of condensed matter research "is to design and rationally prepare complex solids that have predictable and useful properties."[51] But even (or especially) "nearly complete control over the structure of matter" still operates under the constraints of chemical bonding, and models for rationally prepared solids remain mimetically dependent upon naturally occurring molecular configurations. The periodicity of Eigler's atomic IBM logo was determined by the crystalline structure of the nickel substrate on which it was "written," and designs for self-assembling nanostructures (nanotubes, nanocones, nanowires, etc.) evolve from computer simulations of carbon "Fullerenes" and HK97 bacteriophages. So nanotechnology enters into a dialectical collaboration with crystallography and virology, through which nature and culture, imagination and imitation, deviation and constraint are *synthesized* in the pursuit of structural perfection. Richard Smalley, awarded the 1996 Nobel Prize for his "discovery" of fullerenes, articulates the prospect:

> We've got to learn how to build machines, materials, and devices with the ultimate finesse that life has always used: atom by atom, on the same nanometer scale as the machinery in living cells. But now we've got to learn how to extend this now to the dry world. We need to develop nanotechnology both on the wet and dry sides. We need it urgently to get through these next 50 years. It will be a challenge. But, I am confident we will succeed.[52]

Christian Bök's *Crystallography* and Caroline Bergvall's *Goan Atom* are the poems of this climate. Which is not to equate them. As a work that "predicates itself upon an aesthetic of structural perfection,"[53] *Crystallography* is "an act of *lucid writing,* which uses the language of geological science to misread the poetics of rhetorical language."[54] As a work that predicates itself upon the principle that "Anybod's body's a dollmine,"[55] *Goan Atom* uses the "vulgar potential of dropped consonants and arty franglais"[56] to miswrite the perfectionist ethos of genetic engineering. While Bök activates the constraints of axial symmetry, epitaxial accretion, and atomic tesselation as the means of an inspired mimesis, Bergvall turns to surrealist Hans Bellmer's "articulated" dolls and to Dolly, the "sheep," for tropes with which to take on Big Science, working under the rigorous constraint of giving up restraint entirely:

> NO
> workable pussy
> ever was su
> posed to discharge at will
> all over the factory
> sclamation mark (GA, 53)

(Perhaps if "Emily Dickinson" were working today she would write stanzas like that). Following Smalley's taxonomy, one might say that *Crystallography* belongs to "the dry world," while *Goan Atom* is decidedly "wet."

The first two pages of Bök's "Preliminary Survey" indicate the tenor of the matters he takes up:

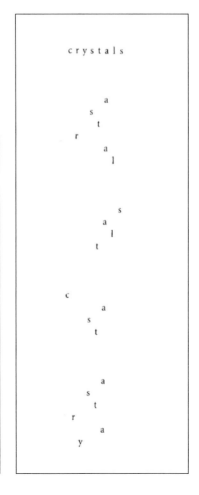

CRYSTALS

A crystal is an atomic tessellation, a tridimensional jigsaw puzzle in which every piece is the same shape.

A crystal assembles itself out of its own constituent disarray: the puzzle puts itself together, each piece falling as though by chance into its correct location.

    A crystal is nothing more
    than a breeze blowing sand
    into the form of a castle
    or a film played backwards
    of a window being smashed.

A compound (word) dissolved in a liquid supercooled under microgravitational conditions precipitates out of solution in (alphabetical) order to form crystals whose structuralistic perfection rivals the beauty of machine-tooled objects.

An archæologist without any mineralogical experience
might easily mistake a crystal
for the artificial product of a precision technology.

A word is a bit of crystal in formation.

(C, 12-13)

"Astral salt cast astray," but traces of "last," "star," "stac," "say," and "art" are also evident. In *Crystallography,* words continually diffract into such casually recombinant "arrays," then tighten into firmer structures:

```
EMERALD

                              O
                              X
                      O    Y
                      OXYGEN
                      O    Y   E
                    OXYGEN
        C             O   Y   E
        H        OXYGEN
        R    S        Y    E
        O    I        G    N
        M    L    E        B              S
    S I L I C O N      B E R Y L L I U M
        U    C    X    B       R            L
        M    OXYGEN   Y   S I L I C O N
             N    G    R    L    I       C    X
             E    Y    L    L         O X Y G E N
             N    L    S I L I C O N    G
                  L        U    C    X       E
        A L U M I N U M    O X Y G E N
                  U             N    G
                  M        O         E
                      O X Y G E N
                  O    Y
                OXYGEN
                  Y    E
              OXYGEN
                  E
        A L U M I N U M
```

crownland beaumontage

sidereal
opulence of sinfulness

opaque, ornate, orphic

silkscreens of silent
orchards

oracular silviculture

alembic of silhouettes

bezels
oblique optics

berylloid observatory

alkali
octane, oxides

ozone overworld of oz

(C, 12-13)

On the left, the chemical structure of emerald generates a lattice of interlocking word-atoms. On the right, we find that these apparent particles are subject to quantum effects, as the first letter of each word has tunneled across the gap between pages, attaching itself to a new semantic unit. Chromium becomes "crownland," Beryllium morphs into "beaumontage," and silicon is suddenly "sidereal." Solid-state chemistry and poetry come face to face in a formal engagement: if "crystals are acrostics generated by the stochastics of a cage" (C, 122) then "a word is a bit of crystal in formation" (C, 12). Acrostics emerging from the chemical structures of amethyst, ruby, emerald, opal, sapphire, jade, and topaz are scattered throughout Bök's collection of crystal information, interspersed between letter-based fractals, a hagiography of snow, a gnomic projection of textuality, a subatomic topography of glossematics, a cryometric index of poetic forms, and the following table of crystal systems for classifying letters of the alphabet on the basis of axial symmetry:

CRYSTAL SYSTEM	SYMMETRY AXIS				CRYSTAL STRUCTURE
	DIA	TRI	TET	HEX	
1. TRICLINIC	–	–	–	–	F G J L P Q R
2. MONOCLINIC	1	–	–	–	A M T U V W, B C D E K, N S Z
3. ORTHORHOMBIC	3	–	–	–	H I
4. ISOMETRIC	(6)	4	(3)	–	
5. TRIGONAL	(3)	1	–	–	Y
6. TETRAGONAL	(4)	–	1	–	O X
7. HEXAGONAL	(6)	–	–	1	–

Unbracketed numerals indicate the mandatory number of axes required for a crystal to occupy a given system. Bracketed numerals indicate the maximum number of optional axes within such a system.

FIGURE 2.8: A table of crystal systems for classifying letters of the alphabet on the basis of axial symmetry

The axis of symmetry describes a mathematical line that passes through the centre of any crystal such that rotation about this line through an arc of 360°/*n* (where *n* ≥ 1) causes the crystal to assume a final position congruent with its initial position. For *n* = 1, the crystal can achieve self-congruence by rotating 360° around an identity axis; for *n* = 2, the crystal can achieve self-congruence by rotating 180° around a diad axis; for *n* = 3, the crystal can achieve self-congruence by rotating 120° around a triad axis; for *n* = 4, the crystal can achieve self-congruence by rotating 90° around a tetrad axis; and for *n* = 6, the crystal can achieve self-congruence by rotating 60° around a hexad axis. No crystal exists with axes of symmetry for *n* = 5 or for *n* > 6.

The number of such axes of symmetry in a given letter determines the crystal system to which the letter belongs: the crystal H, for example, coincides with itself when rotated 180° through any one of three different orientations, and thus the letter belongs to the orthorhombic system, whose members typically have three such diad axes. The alphabet consists predominantly of monoclinic crystals, three types, all having a single diad axis of symmetry: a) three letters symmetrical only through the *x*-axis; b) five letters symmetrical only through the *y*-axis; and c) six letters symmetrical only through the *z*-axis. The triclinic system, the next most common system in the alphabet, contains letters with no axis of symmetry other than their infinite number of identity axes.

The science of crystallography suggests that both the isometric system and the hexagonal system do occur in nature; however, no poet searching throughout the world of language has yet discovered a letter that fits into either system – a mystery that has led some crystallographers to speculate that crystals expressing such a rare degree of symmetry can only exist under the most extreme poetic conditions: perhaps the low temperatures found only in the voids of outer space or the high pressures found only in the cores of neutron stars – conditions difficult for writers to reproduce in the laboratory.

(C, 151)

As a geometric taxonomy of the letter kingdom, Bök's table has its geological counterpart in the "key to speleological formations" that we encounter at the end of the poem "Geodes," the subterranean wonders of which rival Pope's sylph cave. Here, both "p" and "q" are glossed as "precipice with overhanging grotto," while "e" becomes a "grotto with underhanging ledge," and "v" constitutes a "geological fissure cut in bedrock" (C, 60-61). In "Geodes," letters designate the features of a protosemantic topography in which "blindfolds are the logical eyewear" and "handholds are the braille of geology" (C, 46). So we find ourselves back once again at the haptic interface between topography and topology, where (mineral)"sample" meets (finger)"tip," and where atomic structures of linguistic nanocrystals are uncovered that you cannot see, but believe anyway: "A crystal is the flashpoint of a dream intense / enough to purge the eye of its infection, sight" (C, 37).

In *Crystallography,* the cut between vision and tactility constitutes the terrain of the occlusion.

OCCLUSIONS
SUPPOSEDLY
DEPRECIATE
PERFECTION

And as we know,

TRANSPARENCY
IS THE GAUGE
OF ALL VALUE
WHEN CUTTING
WHEN WRITING

(C, 78)

Which explains why scanning tunneling microscopy, for example, is currently being used in studies to "improve rationally the quality of crystals."[57] But despite his investment in "an aesthetic of structural perfection," it's the interface between occlusion and perfection, where topology overwrites topography, in which Bök is ultimately interested:

OCCLUSIONS
PERFECTION

(C, 78)

"Writing," he notes, "represents the superficial damage endured by one surface when inflicting damage upon the surface of another" (C, 124).

Crystallography never gets closer than that to *Goan Atom,* which is occlusion all over. "Sgot / a wides lit," as Bergvall puts it (GA, 23). Presented with the crystal array that opens Bök's book, Bergvall would not fail to include "ars" and "ass" within the range of its combinatory possibilities. Nor would she be insensitive to the scatological implications of "man's" newfound ability to fabricate from the "bottom up."

·		T		s
s		s		o
G		G		F
c		A		A

s		s		·
G		T		s
o		A		A
c		F		G

(GA, 11)

Not unlike Bök's volume, Bergvall's opens with two lettristic grids:

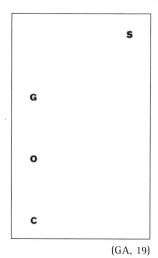
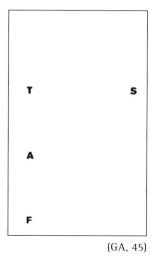
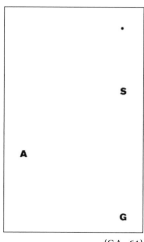

(GA, 19)   (GA, 45)   (GA, 61)

Out of which are drawn the titles for the three long poems that constitute the book:

Along with a decidedly un-astral word string:

> gasp sag toga goat gag cot go cat fag fog tao sat as tag at ass fast (GA, 13)

186

Two pages later, the melo- and logopoeitic satisfactions of this word-search are intruded upon by a phanopoeitic ink blot—one that manages to provide a menstrual counterpart to the anagogic resonance of the atomic IBM logo, Dickinson's proto-graphemic trace, *Carnival*'s dissipative structure, and Bök's crystalline diaspora:

Eventually, we arrive at the appropriate verbal reduction:

(GA, 17)

Blt o by Bolt
Every single P
art is a crown
to Anatom

(GA, 56)

"*Goan Atom*" enlists Goa and anatomy in the anti-literary enterprise of goin' at'em, tirelessly demeaning any Adam's nominative propriety. *Jets-poupée,* the collection's original subtitle, testifies to its mobilization of "French" (without making any excuses) as a dangerous supplement.[58] The most contextually apt translation of *jets-poupée* would be "doll-spray" or "doll-spurt," with *jet* connoting a variety of phallic or mammary prostheses attached to the wet world, including nozzles and hoses. But *jeter* means to throw, and *jetable* means disposable, so the phrase also figures a culture in which not only dolls and body parts, but the sub-ject itself is a commodity—a fetish to which we are hardly faithful, but willing and eager to trade in for other toys. *Poupon*—the masculine noun—means "babe-in-arms." *Poupe*—the feminine noun—means "stern," the two English senses of which are implicitly conflated throughout the book.

The first page of "cogs" is probably the best evidence of the urgency with which Bergvall agitates this crowd of associations and "themes" into a cacophonous poetic riot:

(GA, 21)

Enters the **EVERY HOST**
dragging a badl Eg
Finally !
So that the inspiration for such thoughts
becomes visible through the navel in order
To take advantage of the interior mechanism
run through the thoughts retained of little girls
as a panorama deep in the belly
revealed by multicoloured electric
illumination
it's roped in bottoms up I want to
B ba
b bo
b be
b leed
the load
o
Ff***
Mud and Dead
the mud offal the dead
stuffed goat much
Junk & Gusto
Whats looking at me
Form Form Form
one hardened core after another
bleeding harp
PushPushMarquis
or PunchOlichinelle
Skin-sacks
Minor monticulates
lined Up joind by hind thoughts

OFART
Eat Shit gladly

The EVERY HOST, whose introductory limp presages the staccato cadence of Bergvall's ovular battle, stars alongside DOLLY ("Enters entered"), who is foremost among a whole cast of unseemly characters, including HEADSTURGEONS, FISHMONGRELS, A CO CALLED MOO, and a GROUP OF CORPOREALS possessed of "Fitscrewed Facial sites / Big coily brains," and "throbbing ambient genitals" (GA, 65).

But Bergvall's mention of "minor monticulates" is to the point: a deposition like *Goan Atom* has its oppositional place in the technocultural pornscape evoked by Jean Baudrillard in *The Ecstasy of Communication*:

> There is a fractal demultiplication of the body (of sex, object, desire); seen from up close, all bodies, all faces look alike…The promiscuity of the detail, the zoom-in, takes on a sexual value. The exorbitance of the details attracts us, in addition to the ramification, the serial multiplication of each detail. The extreme opposite of seduction is the extreme promiscuity of pornography, which decomposes bodies into their slightest detail, gestures into their minutest movements. Our desire reaches out to these new kinetic, numeric, fractal, artificial and synthetic images, because they are of the lowest definition.[59]

From this perspective, it is no accident that the "maturation" of nanoscience and technology–channeling its tactile voyeurism into the illusionistic production of sumptuous atomic topographies–is virtually coeval with the explosion of online pornography. *Goan Atom*'s biotechnical obscenities are anything but gratuitous in this milieu, wherein nanoculture mates with pornculture at the interface of "dry" and "wet" worlds. This is the context in which Bergvall's most provocative piece, "Ambient Fish,"[60] must be taken as the definitive commentary on the "nanobots" of K. Eric Drexler's fancy. These, we hear, will soon be "searching out and destroying viruses and cancer cells," "enter[ing] living cells to edit out viral DNA sequences and repair molecular damage." They will "enable the clean, rapid production of an abundance of material goods."[61] "Nanotechnology," enthuses Drexler, "will give us this control, bringing with it possibilities for health, wealth, and capabilities beyond past imaginings."[62] Frankly, this is the kind of money shot that Bergvall has seen before:

Ambient fish fuckflowers bloom in your mouth — will choke your troubles away
Ambient fish fuckflowers bloom in your mouth — will choke your troubles away
Ambient fish fuckflowers bloom in your mouth — will choke your troubles away
Ambient fish fuck flowers bloom in your mouth — will shock your double away
Ambient fish fuck flowers loom in your mouth — will soak your dwelling away
Alien fish fuck fodder loose in your ouch — suck rubble along the way
Alien fish fuck fodder loose in your ouch — suck rubble a long way
Alien fuck fish fad goose in your bouch — suck your oubli away
Alien phock fish fat geese in your bouche — watch a getting a way
Alien phoque fresh fat ease in your touche — watch a ramble away
Alien seal fresh pad easing your touch — take the gamble away
To fish your face in the door
a door a door
fuckflowers bloom in your mouth — will choke your troubles away

Ambient fish fuckflowers bloom in your mouth — will choke your troubles away
Ambient fish fuckflowers bloom in your mouth — will choke your troubles away
Ambient fish fuckflowers bloom in your mouth — will choke your troubles away
Ambient fish fuck flowers bloom in your mouth — will shock your double away
Ambient fish fuck flowers loom in your mouth — will soak your dwelling away
Alien fish fuck fodder loose in your ouch — suck rubble along the way
Alien fish fuck fodder loose in your ouch — suck rubble a long way
Alien fuck fish fad goose in your bouch — suck your oubli away
Alien phock fish fat geese in your bouche — watch a getting a way
Alien phoque fresh fat ease in your touche — watch a ramble away
Alien seal fresh pad easing your touch — take the gamble away
To face your fish in the door
ador ador
fuckflowers bloom in your mouth — choke your troubles away

(GA, 72-73)

Building Dwelling Thinking

Neither *Crystallography* nor *Goan Atom* engage nanotechnology as such, and the "connection" between the poetry I've been reading and the condition of nanoculture cannot be *directly* established, "in the conventional sense of 'positively identified and detained'."[63] But a positivist approach to literature and science studies defeats the purpose of the discipline. Face to face, Dickinson's invisible, pro-creative paragram and McCaffery's cosmo-illogical texture figure the breakdown of spatial conceptuality and classical physics with which nanoscience confronts us. Wread together, Bök's crystal-lines and Bergvall's bio-forms suss out the merger of dry and wet worlds that nanotechnology is accelerating. Altogether, this paper partakes of and belongs to an ever-growing body of scholarship that draws such connections between poetry and technology in order to demonstrate that they were never really separable in the first place.

Ambient fish fuck flowers loom in your mouth	**will soak your dwelling away**

Though this may be something other than the "saying of the poet" that he had in mind, Bergvall's line returns us to Heidegger. "Neither technoscience nor poetry is going to save anything," I wrote. Or rather, poetry may "save lives," but it does not save lives. And science may save lives, but it cannot save life. Still, is it really all that retrograde to ask how we might best go about building, dwelling, and thinking at the limits of fabrication? Nanotechnology popped up in Jacques Derrida's 2003 seminar on "The Beast and the Sovereign" at UC Irvine, as the vanishing linch-pin in a "classic" deconstruction of the concept of scale. The Sovereign, Derrida reminded us, is not only He or That which is higher than height; He or It is also empowered by that which is lower than lowness, smaller than "small." And that magic word, "nanotechnology," was spoken by way of example. The atomic IBM logo almost did not appear in this paper, self-promoting icon as it is of the Sovereign's worst impulses, magnified by "modern technology" at an unbearably intense "resolution."[64] The rhetoric of those in the business of nanoscience and technology is anything but pretty, and we can do ourselves the favor of being sure that its "applications" won't be either. And since we've been reading Bergvall, the acronym "TA" should alert us to the pathos of "technology assessment." When Lacan locates science as drive—not just drive for domination, but species death drive—he means that it operates outside of the ethical domain, despite our best intentions. Those in the Business consistently pepper their calls for funding with calls for "rules and regulations," but those will find their best use in delineating for the public what their corporate government is not "allowed" to do—and therefore does anyway.

But poetry, though it may occasionally point us in the right direction, has no priority as a means of grappling with these utterly intractable problems. Nor is it privileged as a mode of that "sparing and preserving" that Heidegger called dwelling. What is it, or could it be, to "dwell poetically" at the limits of fabrication? Sparing and preserving are not so complicated, but they demand going beyond art and technics and into political commitments, volunteer work, and (after all) those little, nameless, unremembered acts of kindness and of love. Which is not to pose, romantically, another answer, but just to say that at the limits of fabrication, on the *hither side* of complexity, we come face to face with that which is simpler than simplicity. And so instead of extolling a potential "nanopoeisis," I want to close with George Oppen, who gave up poetry for twenty-five years in order to practice..."politics." At the end of "Route," he reminds us of all we know, and all we need to know, on Earth:

> These things at the limits of reason, nothing at the limits
> of dream, the dream merely ends, by this we know it is the real
> That we confront[65]

○○○

Nano Narrative: A Parable from Electronic Literature

JESSICA PRESSMAN

> Human beings have always been digital.
> ◎ Dr. Anders, *Chroma*, Prologue

The first line of Erik Loyer's digital, web-based novel *Chroma* (2001) challenges what it means to be digital by introducing a novel that challenges what it means to be literature. The novel's lead scientist, Dr. Anders, presents an enigmatic statement that collapses human and digital, binary and genetic code, into a correspondence that is both metaphoric and material. Nicholas Negroponte figuratively describes a bit of digital code as "the smallest atomic element in the DNA of information," while Lev Manovich identifies the primary characteristics of new media to be modularity and numerical transcoding—the qualities that enable a correspondence of translation between atoms and bits.[1] Dr. Anders's assertion that biological and digital "have always been" inextricable is an entry point for examining the science of atomic manipulation through literature that manipulates bits. In this essay I construct an analogy between nanoscience and electronic literature, both of which operate through digital mediations to re-present invisible information. Nanoscience is the study and manipulation of matter at the atomic scale, and it depends on digital technologies to transcode, translate, and visualize its interventions.[2] Exploring a digital novel whose narrative resonates with nanoscience, I argue that electronic literature is the literary counterpart to the creation of nano narratives.

My essay relies on other articles in this volume for scaffolding. Sue Lewak's focus on the stimulating power of the imagination, Kate Marshall's argument that nanoscience "emerges as a reflexively-produced technological threat," and Colin Milburn's cogent presentation of "nanowriting" as its own genre support my effort to locate narrative production at the heart nanoscience.[3] I seek to add to these creative critical readings by focusing on the need for visual, digital narrative in nanoscience. Because nanoscience relies on an imagined future (Lewak and Marshall) and the field of literature (Milburn) to construct its identity and chart its progress, it needs narrative—nano narrative. And nano narratives are digital.

Nano Narrative

As the rich tradition of science studies testifies, narrative has long been recognized as a central strategy in creating, analyzing, and disseminating scientific knowledge.[4] The intangible, invisible, and largely incomprehensible nature of atoms renders them especially dependent upon narrative explanation and representation.

As the physicist Ernst Mach explains, "'Atoms cannot be perceived by the senses . . . They can never be seen or touched, and exist only in our imagination. They are things of thought.'"[5] Richard Feynman, the physicist whose famous talk "There's Plenty of Room at the Bottom" (1959) inspired generations of nanoscience research, understood the power of narrative to explain the nano. For example, to illustrate the concept of Brownian motion, Feynman constructed an analogy of people playing with a beach ball and pushing it in different directions; he explained that an observer of Brownian motion "'cannot see the people because we imagine that we are too far away . . . but we can see the ball, and we notice that it moves rather irregularly.'"[6] The beach ball analogy not only visualized the inaccessible phenomenon of molecular motion but also established narrative as a disseminating mode for the science that arose around Feynman's lecture.

Recent advances in the field of nanoscience prompted the Director of the Center for Integrated Nanotechnologies, Terry Michalske, to proclaim "[w]e can really do things we can't imagine right now."[7] Michalske's unnerving statement echoes the facts and fears of technological progress outpacing ethical consideration; it also reverberates with Dr. Anders's chronological collapse, "humans have always been digital," in *Chroma*'s Prologue. Like the novel's preoccupation with communication in the nano-realm of Mnenonos, which I discuss later, Michalske's comment exposes an urgent need for narratives that can imagine and explain the "things" that nanoscience is doing.

Still in the early stages of emergence, nanoscience relies on the literary tradition as a resource for representation. Eric Drexler's *Engines of Creation* introduced nanoscience to the popular imagination by blending science and fiction. He identified literary authors as the imaginations behind nanoscience: "Authors have written of the direct sharing of thoughts and emotions from mind to mind. Nanotechnology seems likely to make possible some form of this by linking neural structures via transducers and electromagnetic signals."[8] Drexler is not alone in utilizing the link between literature and nanoscience. The creators of the Scanning Tunneling Microscope (STM, discussed later), Heinrich Rohrer and Gerd Binning, described its ability to record nano-scale topologies as "getting an image of a river bottom by taking manual depth-soundings—Mark Twain style."[9] The reference to Twain not only locates the "style" of measurement in a technological history but also in a literary one. The STM invented nano writing just as Mark Twain helped invent the American novel.[10] In a recent *LA Weekly* article presenting nanoscience to the general public, Margaret Wertheim writes "like Verne's submarines, STMs plunge us into an enchanted domain beneath the surface of mundane experience."[11] The comparison to Verne is not just a reference but a reliance on the literary imagination whose "willing suspension of disbelief," as Coleridge put it, delineates the invisible and explains the incomprehensible.[12]

Literary language and allusions act as interfaces between nanoscience research and its representation. Stian Grogaard observes, "in the interface between science and its presentation, rhetoric returns."[13] Indeed the words "nanoscience" and "nanotechnology" are rhetorical acts, catchphrases for multiple types of research and fundraising that span the fields of physics, biology, genetics, and engineering. The science of the extremely small employs grandiose rhetoric to construct an identity grounded in literary tradition—specifically, that of the epic quest. The first line of the National Science Foundation's report on the "Societal Implications of Nanoscience and Nanotechnology" (March 2001) reads: "A revolution is occurring in science and technology based on the recently developed ability to measure, manipulate, and organize matter on the nanoscale."[14] Canada's National Institute for Nanotechnology identifies nanoscience research as "opening up vast new horizons," and the first line of Edward Timp's introduction to *Nanotechnology* connects Marco Polo's "venture beyond the horizon" to "explorers of a new frontier; a frontier that exists on the head of pin [sic]."[15] Although its rhetoric is epic in scope, nanoscience's literary interface is media-specific. The frontiers of invisible information are accessible to human perception only through mediation by digital reading machines. Thus the literary interface for nanoscience is the computer screen and the digital novel.

Atoms, Bits, and Electron.ic Literature

Although the existence of atoms had been theorized since ancient Greece, direct access to the atomic realm has only recently been made possible through microscopes dependent upon digital (as well as analog) technologies for imaging and visualization. In 1959 Feynman complained, "our mechanical computers are too big; the elements in this box are microscopic. I want to make some [computers] that are submicroscopic."[16] Advances in the miniaturization of electronic components allowed molecular matter to be transcoded into binary code, thereby transforming nano-space into cyberspace.

As Friedrich Kittler states, "Media determine our situation" and consequently participate in shaping our reality and the narratives we tell to explain it.[17] Marshall McLuhan's understanding of technology as an extension of our bodies aptly describes the machine used to read atomic matter. The Scanning Tunneling Microscope (STM) transmits topographical information about an atomic surface by turning touch into image, "reading" by "feeling" the atomic surface.[18] The STM transcodes tunneling currents, electric energy, into quantified information and transfers this data across analog and digital platforms.[19] At the atomic scale, matter is not static; rather, one researcher compared the forces working at the atomic surface to being in gale-force winds on the macroscale. Using quantum interactions, the STM reads and records changes in the tunneling currents that correlate with variations in the atomic surface. To make this data accessible to human intuition, visualizations are created. The visual image produced by the STM is thus not a direct representation of a stable object but a *description* of dynamic interactions between quantum forces. The results are narratives that inscribe the invisible and capture the kinetic.

Electronic literature is also a re-presentation of electrical action. Kittler writes, "All code operations, despite their metaphoric faculties ... come down to ... signifiers of voltage differences."[20] Created, accessed, and distributed on the computer, electronic literature is digital in its artistic conception and creation, as well as its distribution and reception. This definition depends not on the distinction between print and digital media but on the computer as a reading machine that *produces, controls* and *manipulates* digital bits in order for the literary work to emerge and be experienced. Electronic literature can thus be seen as a literary counterpart to nanoscience. Like the firing electrons that motivate the energy state of an atom, "the code is what makes [digital text] flicker," John Cayley observes, "what transforms them...into writing as the presentation of *atoms* of signification which are themselves time-based" (emphasis added).[21] Like the STM's transcription of the atomic surface, digital programming for electronic literature is also a description: of action and appearance, software specifications, and temporal causalities. Electronic literature *happens;* it is not the code as object but the event as emergence that constitutes the work.[22] Operating through a series of encoded semiotic translations across programmable and binary languages as well as between the CPU, controlling hardware, software, and peripheral devices, electronic text "cannot simply be seen as something which goes on behind the screen; it emerges when....the composed code runs," as Cayley writes.[23] As the circuitous layers of digital interactions occur, the programming plays, and the "electronic text exists as a distributed phenomenon," as Katherine Hayles reminds us.[24] Electronic literature therein subverts the traditional ability to locate a static textual object or signifier, performing Derrida's idea that "[p]lay is the disruption of presence."[25]

In the networked operations that "play" electronic literature, "the boundaries between the text and the context have begun to dissolve," as media critic Peter Lunenfeld articulates.[26] This collapse is in some ways characteristic of digital media. Pierre Lévy substitutes the term "unimedia" for "multimedia" to express the fact that, as Lunenfeld explains, "the computer is the universal solvent into which all difference of media dissolves into a pulsing stream of bits and bytes."[27] Following a similar line of thought, Rita Raley observes that the electronic literature is a "performance [that] collapses processing and product, ends and means, input and output, within a system of 'making' that is both complex and emergent."[28] Electronic literature is a site-specific production in which the configurations of the user's computer set the stage for a literary performance and "[t]he interface becomes the arena for the performance of some task in which both human and computer have a role," as Brenda Laurel observes.[29] Understanding that electronic literature is a visual-verbal happening closer to performance than print, rips open the traditional framework of approaching and analyzing literature. The challenge posed to readers and critics of electronic literature is to avoid the "categorical error," as Lunenfeld calls it, of applying the "aesthetics particular to the static object" to "the dynamic arts."[30] Taking a cue from nanoscience may help us to retain composure as we pursue analysis of the digital performance that is electronic literature.

Since atoms are constantly in motion, interacting with each other in environmentally specific systems, the reading of a single atom is an extraction. The Heisenberg Uncertainty Principle, a central theory of quantum mechanics that informs nanoscience, states (in simplified terms) that the position and momentum of a subatomic particle cannot be measured simultaneously beyond the limits of the Planck constant. Measuring a single component of a larger, dynamic system is an extraction, a recognition that informs our reading of atomic images as well as electronic literature.[31] Practicing a traditional (print-based) approach to reading digital literature, such as focusing only on written text is analogous to attempting to read a single atom—freezing time (literally, since atomic motion slows at cooler temperatures), extracting an interactive element, and representing it as a bit (again, literally) of discrete and, now, discursive data. Rita Raley uses the analogy of Jasper Johns's painting *Flags* (1967) to illustrate the anamorphic character of electronic literature. Just as one of the two flags in John's painting must recede for other to emerge, in electronic literature, she writes "[m]eaning happens in the exchange, but the exchange can never be fixed—it just happens."[32] Understanding the connection between atomic behavior, Heisenberg's Uncertainty Principle, Jasper Johns's *Flags,* and electronic literature prompts us to reconsider our approaches and expectations to reading both the organic and the digital.

Erik Loyer's *Chroma:* A Parable for Nanoscience

Chroma is a digital novel that embraces and exposes the challenges and transformations of electronic literature within a parable for nanoscience. *Chroma* resonates with the ambitions and themes of nanoscience although it never directly names its affiliation. The novel narrates the quest to access and control the mnemonos, an organic realm filled with an invisible information-carrying matter called "marrow." Dr. Anders, the leader of the expedition, describes the mnemonos as:

> A natural cyberspace where the things of the mind appear
> as real as anything your five senses perceive.
> This realm is filled with a substance called "marrow"
> which transmits information just as the air transmits sound. (Prologue).

A "natural cyberspace" is an oxymoron, but one that resonates with nanotech-related fields like "molecular engineering" and "biotechnology." Reality is different at the nano scale. Susan Strehle explains "[s]ubatomic reality is energy, rather than matter, for particles whose position and velocity cannot be determined cannot be said to 'exist' as things."[33] As the identity of matter changes so do the rules governing material reality. Feynman explained that atoms "on a small scale behave like *nothing* on a large scale," and therefore "we are working with different laws."[34] Strehle writes, at "the subatomic level, reality is discontinuous or quantized."[35] Nanospace is a natural cyberspace; it resembles the digital form of the novel and *Chroma*'s description of the "natural cyberspace" called "mnemonos."

The mnemonos is named for the Greek goddess of memory, Mnemosyne, who is also mother to the Muses and inventor of words. W.J.T. Mitchell writes that memory "is the mental power which preserves and orders the phenomena of experienced time."[36] Memory is thus intrinsic to creating plot or narrative. Cognitive theorist and artificial intelligence researcher Roger Schank argues that memory is actually structured and enabled by narrative.[37] To name as the "mnemonos" the nano-like realm that stimulates the imaginations of Dr. Anders and his crew is to position narrative at the center of their explorations into an organic and invisible space. Describing the marrow-filled mnemonos, Orion17, one of the three researchers called "marrow monkeys," describes the "natural cyberspace" in language that resounds with nanoscience and its biotechnological convergence of human and digital. Orion17 relates that the mnenonos is a place in which "we could create and recreate the *tiny machines*/ working away in our psyches,/ giving them form, and physics" (Ch. 2, emphasis added). His dream of creating such "tiny machines" echoes Drexler's early visions of "tiny assemblers," molecular machines called "grey goo." Orion's nanotechnological depiction of the machines we create reflexively constructing our psyches and our memories points to the recursivity of the posthuman circuit in which humans are cyborgs informed, or "situated," by our technologies.[38]

Although the novel is infused and informed by a posthuman understanding of technological recursivity, technology is conspicuously absent from *Chroma*. Dr. Anders's prologue introduces the mnemonos as an Eden, from which our expulsion inspires that "every electronic device,/ every computing machine,/ is an expression of yearning/ to return" (Prologue). The novel references cell phones, computer programming, digital languages, and electronic machines but never depicts their use nor reveals the type of technologies employed by the expedition. However, *Chroma*'s characters are all identified, with the possible exception of Dr. Anders, by Internet usernames: Orion17, Duck at the Door, Grid Farmer Perry. They are already logged into the digital network, interpolated into a recursive system in which the marrow monkeys exhibit Katherine Hayles's understanding of posthuman subjectivity in which "[c]onsciousness alone is no longer the relevant frame but rather consciousness fused with technologies of inscription."[39] From its first line, *Chroma* presents the congruity of human and digital, and the transcendental cyberspace is filled with a substance named for the most bodily and vital in the human skeleton—marrow. *Chroma*'s posthuman "fusion" of organic and digital renders that the realm of memory and narrative is digital Consciousness and technology merge in the "natural cyberspace" of the novel and the actual cyberspace that contains it.

Narrative is at the heart of nanoscience, and at the heart of *Chroma*'s narrative is nanoscience. Believing that he has "found the key/ that can unlock the ancient realm" of the mnemonos, Dr. Anders recruits three researchers, called "marrow monkeys," to "explore the inner workings of the mnemonos,/ and then help me to safely reveal its intricacies to the public" (Prologue). The last part of this mission statement is the most important: for Dr. Anders, the marrow monkeys, and my presentation of *Chroma* as a parable for nanoscience. It is not the investigation of the mnemonos that compels the scientific expedition but its *narration*. *Chroma* thus presents narrative as the provocation, process, and product for the novel's metonymic nanoscience.

Like the atomic realm, the mnenonos defies direct human interaction; it is a different reality.[40] Dr. Anders realizes that the success of the scientific venture depends on the ability to transform personal experience into communication, to not only access the mnenonos but to narrate it. When communication fails, Dr. Anders has nightmares of a Tower of Babel situation, "Visions of some hell in which/ their self-absorption continues until/ none of us share a language in common" (Ch. 3). Orion17 explains their error: "We simply assumed that this world's marrow/ would somehow provide instantaneous translation/ between our myriad intimate tongues" (Ch. 2). The issue is one of synchronicity: a failure of transmission and translation, an incongruence of information and interaction. Depicting a scientific expedition into a nano-like realm, *Chroma* exposes a gap between science and its representation. The novel illuminates the layers of mediation and translation inherent in a science, like nanoscience, that is inextricably bound to its instruments and interfaces. To make their research relevant, to "reveal" their experiences to the public, the marrow monkeys must find a way to cross platforms and physical places, marrow and mental paradigms, in order to produce and disseminate their narratives about the mnemonos.

To combat their inability to communicate in the mnemonos, Orion devises a plan: the marrow monkeys will construct avatars that will enable them to interact with each other in the natural cyberspace. Orion17 explains, "I've been thinking about our own DNA,/ our genetic code, as a kind of program/that *describes* a human body. What if we could write programs that would make/ bodies for us in marrow?.... We'll design avatars, virtual bodies for us/ to wear while we're in marrow" (Ch. 5, emphasis added). Orion17's plan is based on the perception of genetic code as narrative, and it hinges the expedition's success on the ability to write narratives. He continues, "If we agree in advance on exactly how these bodies/will function, and exactly what they can look like, /I think we'll finally be able to see each other, /talk to each other, while we're in the mnemonos" (Ch. 5). Like novelists, the marrow monkeys must conceive of their avatars as characters in autobiographical stories. Although their creation of digital avatars replicates the genetic production of bodies, it is not the body as physical object that matters but the *narrative description* of it.

Chroma's final chapter, however, disallows a privileging of the discursive or a utilitarian view of the avatar as it complicates the collapse of binary and genetic code. In chapter six Duck at the Door depicts the creation of an avatar within a racial, cultural, political framework. Duck is female and bi-racial, a hybrid "invented body" (Ch.6). She thus understands information and interaction to be embodied and intertwined in social and cultural "games" of signification. On her "character record,/ right below the words 'Choose only one,'/ the African-American and Caucasian boxes/ are both checked," a fact that identifies Duck's body as a challenge to the "rules of the game" of social structure (Ch. 6). In her "real" life (outside of the mnemonos), Duck's experience is similar to those the problems plaguing the marrow monkeys while in marrow: "a lot of people don't see me at all" (Ch. 6). Duck's chapter disables utopian claims about "disembodied" realms like the mnemonos, nanospace, or the World Wide Web. She writes, "Unsettling things were always happening to me...Unjustified love,/ unjustified hate,/constant misunderstandings,/and all because of my invented body,/my avatar" (Ch. 6). Chroma's final chapter reminds us that our choice and use of mediating technologies always informs our readings, interactions, and the narratives we use to explain our world.

Literature Becomes Electronic…

As an online digital novel, *Chroma* enacts Duck's (and Kittler's) understanding that our media mold our realities while it illuminates how literature changes when it becomes or (according to Dr. Anders) has "always been" electronic. *Chroma* is structured as a traditional novel, with chronologically numbered chapters that contain journal entries by the characters. But disturbing the genre expectations is the fact that *Chroma* can be accessed through two reading registers—View Text and Perform Text—that contain the same narrative content but enact very different presentations.[41] The registers cannot be read simultaneously, for they demand separate windows that cover the screen. The reader must therefore select her method of mediation.

Perform Text is the default option for experiencing *Chroma*. This register subverts the hierarchical privilege of text in literature for it contains minimal, if any, written text. There are no hyperlinks or interactive choices, and there is no option to pause the frame (other than closing the chapter window entirely). Instead, a Shockwave animation proceeds continuously with design dancing in synchronization to music and voice-over narration. The following image is a screenshot, but in the actual chapter, the reader witnesses an animated performance of graphic, sonic, and, dynamic narrative modes.

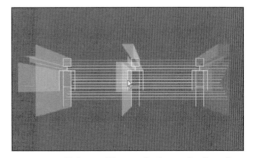

Perform Text moves across the screen and the senses, provoking questions about what "text" is in electronic literature.

Erik Loyer *Chroma*, Prologue, Perform Text

197

If Perform Text is *Chroma*'s default option, View Text is the remix.[42] View Text presents written text in a vertical register with the same music from Perform Text but without animation or voice-over narration.

The View Text section resembles the traditional mode of reading online: the reader uses her mouse to direct the vertical scrollbar and encounter text in stanza form. However, View Text resists expectations and instigates questions about what it means to "read" electronic literature. The section contains written text that

Erik Loyer *Chroma*, Prologue, Perform Text

defiantly proclaims its independence from inscription; individual letters flicker boldly in discrete isolation from the rest of the word or line. This visual reference to the "flickering signifier" reminds the reader that digital textuality demands and incites a distinct type of reading.[43] The reader engages with atomized text that is defiantly digital and experiences Michael Joyce's observation that "electronic text is the constantly replaced present tense."[44]

As literature becomes electronic, the method and meaning of reading changes. The View Text music orchestrates a reading rhythm, propelling eye movement and speed, as the reader becomes aware that the novel and computer are in collusion to prevent her control. The scrollbar on the text panel appears to be the common instrument for navigating web pages, but *Chroma*'s scrollbar is actually programmed to skip between stanzas and disallow scrolling by line. Eliding direct manipulation, the misleading navigation tool promotes an awareness of the physicality involved in navigating and reading electronic literature. View Text engages the reader in a performance with flickering text in order to generate a literary happening that transforms the reading subject.

Electronic literature and nanoscience are emergent narratives that inform, transform, and express our posthuman understanding of the world and our methods of reading and narrating it. Translating data across informational modes and media through cyborgic extensions and avatar representations, the digital computer and the STM are both part of the "discourse network," as Kittler calls it, "that *allow[s]* a given culture to select, store, and produce relevant data" (emphasis added).[45] *Chroma* engages and exposes the digital discourse networks that transform our reading practice and the readings we produce. Enacting an anamorphically informed reading process of moving between Chroma's two registers, we become aware of the mediating technologies that extend our reach into "natural" and digital cyberspaces and *allow* the stories we tell about them.

An "Unfinished" Conclusion

Chroma is an unfinished work, and it relishes this state. Intended as the second of a three-part series, the novel currently contains only seven of the planned sixteen chapters.[46] Its unfinished nature resonates with Jane Yellowlees Douglas's observation that digital stories lack closure and Michael Joyce's dictum that "every story goes on without us."[47]

The final screen is "titled" "To be Continued?" (Ch. 7).[48] There is no music, animation, or narration. Instead, this screen addresses the reader directly to explain that for *Chroma* funding has run out. *Chroma* exists within the commercial network that is the Internet, and its last screen contains details about materials of production.

To be continued?

Thanks for visiting Chroma. Due to the generous support of The Rockefeller Foundation, Chapters 0-6 of this project have been completed.

However, while the script, design and voice-over for the remaining chapters is done, there currently remains about a year and a half of labor involved for me to be able finish the project on my own.

While production on Chroma has currently been suspended, I am seeking additional support to fund the completion of the remaining chapters. Also, if you're a skilled dialogue editor or composer willing to volunteer your time to produce elements for one or more chapters, I'd love to talk to you.

If you are interested in supporting the future development of Chroma and want to learn more about the balance of the project (what's been seen to date is really the first act of a three-act drama), please contact me, Erik Loyer, at erik@marrowmonkey.com.

In addition, I am also looking into producing a limited edition CD-ROM containing the complete script of Chroma, all the finished episodes, behind-the-scenes material and an archive of past work.

Subscribe to my newsletter to be kept up to date on the CD-ROM and all things Chroma, as well as my other projects.

And again, thanks for tuning in.

-Erik Loyer
December 2002

Erik Loyer *Chroma*, Prologue, Perform Text

The inclusion of a narrative about financing (a grant from the Rockefeller Foundation) and creative practices (an invitation for collaboration) complicates the aesthetic closure of the novel. The final screen also contains active hyperlinks to email the author or subscribe to a *Chroma* newsletter. The novel thus reaches beyond its narrated story about a "natural cyberspace" to enter the "real" cyberspace of the World Wide Web. Performing Peter Lunenfeld's aesthetic of "unfinish," *Chroma* supports his claim that "the limits of what constitutes the story proper are never to be as clear again."[49] Susan Sterhle writes that the "new physics has reimagined reality."[50] So too has digital technology reimagined literature. The challenge to the reader, critic, and creator of electronic literature—the "explorers of a new frontier" (to use nanoscience rhetoric)—is to re-imagine our assumptions and approaches to understanding, analyzing, and enjoying literature.

○○○

What's the buzz? Tell me what's a-happening[1]:
Wonder, nanotechnology, and *Alice's Adventures in Wonderland*
SUSAN E. LEWAK

> The most beautiful thing we can experience is the mysterious. It is the source of all true
> art and science. He to whom this emotion is a stranger, who can no longer pause to
> wonder and stand rapt in awe, is as good as dead: his eyes are closed.
> ⊚Albert Einstein

> Once upon a time, there was a princess who lived in a castle in a faraway land. One day,
> she and her brothers and sisters escaped the evil king who held them captive there and
> tried to make their way in the world. It was hard because the king never stopped looking
> for them...One day, the evil king and his men found her and took her away, so there was
> no happily ever after.[2]

These are the final words of "Episode 19: Hit A Sista Back," from *The Complete First Season* of James
Cameron and Charles H. Eglee's *Dark Angel*. This short-lived cyberpunk television series tells the
story of Max Guevera (X5-452) a "genetically revved up female," who was created from feline and
human DNA in the secret government lab, Manticore. Episode #19, however, focuses on one of Max's
sisters, Tinga, who along with Max and ten other transgenic nine-year-olds, escapes from the "castle"
in the year 2009. Ten years later, Tinga is forced to return to the "castle" by the "evil king" (government agent Donald Lydecker) in order to save her son, Case, who has been infected with a fatal virus
by Lydecker's team of biotechnical engineers.

Within the paradigm of science fiction, where the transgenic superhuman is accepted as an alternate
form of "the real," this plot appears familiar. It becomes unfamiliar, strange, and wondrous, however,
when Max asks Lydecker if the disease Tinga's son is suffering from is due to "some kind of bioengi-
neered virus." Lydecker's single crisp affirmation, *"Nanotechnology,"* suggests not only an evil force
but also one which neither Max nor Tinga can defeat.[3] The further positioning of this episode as an
anti-"fairy tale" in the lines quoted above suggests that if transgenic super humans are "the real," the
realm of wonder or magic lies within the still mysterious field of nanotechnology.[4]

Why are fear and wonder so often linked with nanotechnology? After all, the concept of the
nanoscale (one nanometer is one-billionth of a meter) is far from new in the scientific community.
Since the late twentieth century, however, the term nanotechnology has elicited an explosion of
interest in the greater public and private sector. This is happening, despite the fact that, as Michael
L. Roukes reminds us, nanotechnology has yet to mature as a science and "is still largely a vision for
the future."[5]

What's the "buzz" about, then? And is it related to the power which wonder holds over us? Wonder in
the context of nanotechnology denotes mystery: a state of unpredictability and chaos which implies
that at this miniature level, everything and anything is possible. In fact this assumption is, like a circus

magic act, founded in illusion. Rather than outright chaos, laws that are currently under exploration but not entirely understood govern the nanoscale. With time, as our understanding of these laws increase (thus diminishing the state of wonder) will the "buzz" remain? Or—as with the *Space Age* of the fifties and sixties or the *Dot-Boom* of the nineties—will interest wane as wonder becomes ordinary?

If Lewis Carroll's 1865 text, *Alice's Adventures in Wonderland* (perhaps aptly described as the "Harry Potter" of the mid-nineteenth century) is any indication, it is the idea of wonder itself, rather than the entity associated with wonder, which causes the excitement.[6] Alice's journey, which the *National Science Foundation at the University of California* and others claim as a useful metaphor for nano-technology, raises a potential answer to the question: "What's the buzz?"[7]

Initially, the buzz appears to be about the technology itself that, if realized, promises to remove us from the binding nature of the familiar. However, Alice's gradual loss of interest in Wonderland as she real-izes that it is organized through a different (rather than magical) set of rules is important to note. This realization is the culmination of three stages of development: a) the illusion of chaos due to a lack of understanding of the rules (Chapters I-V); b) the gradual acclimation and adaptation to the new set of rules (Chapters VI-X); c) the mastery of the rules and consequent "awakening" from wonder (Chapters XI-XII). Likewise, once the current discussion of the nanoscale is replaced by a new understanding of its laws, once wonder becomes familiar, the "buzz" may dissipate due to a similar awakening.

As a recent headline in the *San Francisco Chronicle* indicates "Silicon Valley pins hopes on nano-technology boom; U.S. ready to spend billions on revolutionary science."[8] When exploring the origin of this "buzz," a logical beginning place is the counterintuitive nature of working at the nanoscale, where behavior among atomic particles narrows old expectations. As long as these expectations remain, the nanoscale appears to lie within the realm of the fantastic (and thus appeals to our desire for freedom from physical laws). Yet, as Michael Roukes indicates: "this domain is not some ultra miniature version of the Wild West. Not everything goes down there; there are laws."[9]

The inability to detect these laws, however, often invokes a sense of wonder among those who are interested in the workings of the nanoscale. Or, as Jessica Pressman notes, "nanoscience thus depends on narrative."[10] Brooks Landon further demonstrates the tie between nanotechnology and the percep-tion of magic by noting that:

> While it is easy to criticize the numerous nanotechnology narratives that unrigorously attri-bute magical properties to nanotech (Ian Watson's *Nanoware Time* comes to mind), it's hard to resist thinking of as magical a technology that envisions disassembling any and all mat-ter by breaking it down into constituent molecules and atoms and reassembling those mol-ecules and atoms into an unlimited range of substances, objects, and even human beings.[11]

Thus Colin Milburn's observation that "nanotechnology is science fiction" underscores the assump-tion that at the nanolevel, the impossible becomes the real.[12]

What differentiates nanotechnology from other forms of science or science fiction? What's the "buzz" about? The answer may lie within the very components of the "fairy tale." As Steven Swann Jones notes, the fairy tale is merely one type of folktale, along with fables, jokes, and novellas. While these forms of folk narrative employ the use of ordinary protagonists (as opposed to myth which is limited to tales of the Gods), the fairy tale is differentiated not by the use of "fairies" (which do not necessarily appear in the tale) but by magic: "While [fables, jokes and novellas] are reasonably mimetic—that is, they depict life in fairly realistic terms—fairy tales depict magical or marvelous events or phenomena as a valid part of human experience."[13] The excitement of nanotechnology thus lies in its resemblance to a fairy tale: magic that is predicted to become a "valid part of human experience." As it already exists in the future, nanotechnology appears as a wondrous "reality" because it WILL BE, not because it IS. Or as Kate Marshall notes, nanotechnology's present conception is determined by its anticipated future."[14]

Carroll's revolutionary vision of wonder made *Alice's Adventures in Wonderland* a unique contribution to the fairy tale revival in England during the 19th century. Previously, folktales in the oral tradition were hardly suited for children. Indeed, as Jack Zipes indicates in his discussion of the *Little Red Riding Hood* cycle, the tale in its oral form verged on the taboo as it "reflected the sexual frankness of the peasantry during the late Middle Ages."[15] It was largely due to the literary endeavors of Charles Perrault during the 17th century and the Grimm Brothers during the 19th century that the oral tradition mutated into didactic morality tales for children.[16] As Peter Hunt indicates, however, *Alice's Adventures in Wonderland* was important for its attack against this movement and is," generally regarded as the greatest turning point in nineteenth-century children's fiction, a book where children's imaginations were first given absolutely free reign, with no moral messages."[17]

While Carroll's rejection of the didactic nature children's literature may not sound revolutionary to a contemporary audience, it was liberating for 19th century readers restricted to a diet of morality-driven fantasy. It is this vision of wonder which links *Alice's Adventures in Wonderland* to nanotechnology: although wonder appears in each to be unlimited (and therefore liberating), both Wonderland and the nanoscale are actually governed by sets of rules.[18]

The first of the two "Alice" books was originally conceived of as a hand-written and illustrated gift (entitled *Alice's Adventures Underground)* for Carroll's favorite "child – friend, "seven-year-old Alice Liddell.[19] It was only after strong encouragement that *Alice's Adventures in Wonderland* was published in 1865 for commercial distribution and illustrated by the noted artist, Sir John Tenniel. Likewise, though Carroll had not intended to write a sequel, strong persuasion from publishers and fans of Alice led to *Through the Looking Glass* and *What Alice Found There* (1871). As a realm of fantastic reversals rather than microscale "nonsense," the world which Alice encounters beyond the looking glass is distinctly separate from Wonderland. Despite this fact, the two novels are usually published together as the single text, *Alice in Wonderland.*

Alice's Adventures in Wonderland tells the story of a seven-year-old curious girl who follows a talking white rabbit down a surreal rabbit hole which leads to a wondrous and fantastic land, only accessible at the microlevel. Though initially flustered by the deceptively chaotic nature of the land, Alice quickly learns that there is a "logic" to Wonderland, but one which requires her to discard former expectations. Indeed, it is important to note that neither Wonderland nor the realm of nanotechnology are neat inversions of our macroscale logic. If Alice later discovers that the world "through the looking glass" is a mirror reversal of the macroscale, she must equally learn to accept Wonderland's microscale logic as a separate entity which operates according to its own set of principles. Brian

Attebery notes that this is a primary characteristic of nanotechnology as: "much nonsense, though, becomes demonstrable sense at the submicroscopic scale...the world described by quantum physics is...a weird wonderland where common sense is just plain wrong."[20]

As Alice will discover, these "terms" center on the notion of scale. If macroworld logic is dependent upon a linear notion of size and growth (a child grows to become an adult), microworld logic is reflected through the inability of Alice's body to maintain one particular size for a period of time. Alice is thus confused only so long as she clings to a linear notion of scale. Once she adjusts to the new multi-linear nature of her body (i.e. its inability to maintain a static size) she will be able to accept microworld logic as valid. She will thus be able to understand Wonderland not as "wondrous" but simply as an "IS" which cannot be explained through macroworld expectations.

Alice's experiences thus replicate those of the visitors to LACMALab's *nano* as described by Adriana de Souza e Silva at the beginning of this volume. One of the fundamental objectives of the exhibit is to use macrolevel expectations to explore microlevel "realities." Visitors to the *nano* exhibit will attempt to interpret their experience through macroscale logic, an attempt which will ultimately fail. The goal is thus to help visitors use this failure as a "rabbit hole" down to an entirely new set of expectations. For example, when entering the *Sense Space*, Silva states: "your idea of what is stable and physical is endangered, because you always thought the most accurate way of perceiving the world was by seeing. Immediately you remember the STM (Scanning Tunneling Microscope), which tells us that the nano-world cannot be viewed, only felt."[21] Visitors must learn to understand that the nanoscale is a realm in its own right, with its own set of rules rather than a simple inversion of macrolevel logic.

This development in understanding is demonstrated through Alice's inability to recite poetry properly. As will be discussed in detail later, Alice's attempts to recite upon demand produces poems which mimic the rhythm and beat of the work, but disrupt and distort the original linguistic content. She thus discovers that it is impossible to use macroscale language at the microscale. Niels Bohr made a similar determination about the activities at the quantum scale (which govern the nanoscale). He argued that interpretations of activities at the quantum level are irrevocably limited by the method of interpretation as "theoretical concepts are defined by the circumstances required for their measure-ment."[22] Behavior at the quantum level thus only appears to be wondrous: it is, instead, the inad-equacy of language as a means of interpretation which leads to this conclusion.

Bohr's argument leads to a new relationship between subject and object as indicated by Karen Barad when she argues that:

> [Agential Realism] is a theory of knowledge and reality whose fundamental premise is that reality consists of phenomena that are reconstituted in *intra-action* with the inter-ventions of knowers. 'Intra-action' signifies a dynamic involving the inseparability of the objects and agencies of intervention (as opposed to interactions which reinscribe the contested dichotomy).[23]

If macroscale language implies that there will be a natural separation between subject and object, microscale language conflates the observer with the observed so that one cannot be understood as separate from the other. Furthermore, if in the macroworld, Alice could "recite properly" and depend upon a body that was relatively static, at the microlevel (where Alice absorbs and becomes part of the language of nonsense) she cannot recite properly and her body is perpetually reconfigured. The "magic" of the nanolevel is thus a linguistic one: with the conflation of the subject and object, the macrolevel lacks sufficient means to interpret microlevel behavior.

Alice initially perceives Wonderland (Chapters I-V) to be a chaotic and disorderly place. This is due, in part, to its existence outside of macroworld expectations (as represented through the linear nature of print):

> Alice was beginning to get tired of sitting by her sister on the bank and of having noth-
> ing to do: once or twice she had peeped into the book her sister was reading, but it had
> no pictures or conversations in it, 'and what is the use of a book,' thought Alice, 'without
> pictures or conversations.'[24]

If her sister's book is emblematic of a linear (and therefore non-wondrous) environment, Alice's imaginative desires (the multi-linear Wonderland) indicate that wonder can only exist at the micro-level. In order to do so, she assumes the role of the "Weaving Woman," a popular motif (within folk-lore) for the female storyteller: "So she was considering, in her own mind (as well as she could for the hot day made her feel sleepy and stupid), whether the pleasure of making a daisy-chain would be worth the trouble of getting up and picking the daisies, when suddenly a White Rabbit with pink eyes ran close by her" (7).

The first indication of Wonderland's chaotic appearance occurs during the long fall down the rabbit, an act which removes Alice from macroworld expectations of time and space: "Either the well was very deep, or she fell very slowly, for she had plenty of time as she went down to look about her, and to wonder what was going to happen next . . .down, down, down. Would the fall never come to an end" (8)? Now deeply within the realm of wonder, the multi-linear worlds of the visual and aural once dominate in extraordinary ways. As she falls, Alice notices not the brown dirt of a rabbit hole but instead cupboards and shelves filled with jars of marmalade. The power of the microworld has already taken effect: rather than display fear, Alice is bored and begins to speak the wondrous language of no-sense: "There was nothing else to do, so Alice soon began talking again. . . .'But do cats eat bats, I wonder?. . . .Do cats eat bats? Do cats eat bats?. . .Do bats eat cats?' for, you see, as she couldn't answer either question, it didn't matter which way she put it" (9).

The second indication of Wonderland's potential for chaos occurs when Alice initially confronts the question of scale: wonder is only accessible at the microlevel, as indicated by a long hallway of miniature locked doors. Although she tries, at this juncture, to analyze her options in an orderly macroworld fashion, her attempt is in vain. Alice has already been corrupted by the presence of microworld logic, as indicated by the internal debate she has over the wisdom of drinking liquid from a mysteriously labeled bottle:

> It was all very well to say 'Drink Me' but the wise little Alice was not going to do that
> in a hurry. 'No, I'll look first,' she said, 'and see whether it's marked 'poison' or not'; for
> she had read several nice little stories about children who had got burnt, and eaten up
> by wild beasts, and other unpleasant things, all because they would not remember the
> simple rules their friends had taught them (10-11).

Finally, the question of identity as well as the prominence of the visual both serve to convey the co-existence of chaos and "order" within Wonderland. After drinking the potion in the bottle and shrinking to the microlevel, Alice finds herself incapable of correctly answering the Caterpillar's question, "Who are you?" (35). Macroworld identity is thus tied to linear notions of the body as Alice indicates: "I-I hardly know, Sir, just at present–at least I know who I was when I got up this morning, but I think I must have been changed several times since then . . . being so many different sizes in a day is very confusing" (35). In addition, Alice's loss of identity is tied to a new-found inability to recite poetry (which one assumes she was able to do in macroworld classrooms) as she states "I can't [sic] remember things as I used" (36). Her attempt to perform, upon the demand of the Caterpillar, the poem "Old Father William" is notable for two reasons: a) her inability to recite it properly, b) the presence of highly detailed illustrations with each verse. Be careful what you ask for as you may be granted your wish: Alice's desire for a visual and aural world is fully granted at the expense of a macroworld sense of self and identity.

Alice's new relationship with the notion of scale, however, also indicates the existence of a new set of rules at the microlevel. After drinking the potion, Alice states: " 'What a curious feeling!. . . I must be shutting up like a telescope!' And so it was indeed: she was now only ten inches high" (11). This descent to the microlevel, however, is not only described in the text above, but also conveyed to the reader through rows of asterisks which replace what should be three to four lines of text:

```
    *       *       *       *

    *    *      *

    *       *       *       *
```

The precision of asterisks, not randomly placed but rather geometrically aligned, coupled with the newfound instability granted to Alice's body indicates substitution: the replacement of macroworld linear sense with the visual and multilinear microworld *no-sense*.

Meaning at the microlevel therefore does occur through the granting of Alice's wish for a story with pictures and conversations. However, it is a new notion of "sense," one which acts according to its own (rather than the macrolevel's) set of rules. This is further evidenced through "The Mouse's Tale" in Chapter 3:

"Fury said to
a mouse, That
he met in the
house, 'Let
us both go
to law: I
will prose-
cute *you*.—
Come, I'll
take no de-
nial: We
must have
the trial;
For really
this morn-
ing I've
nothing
to do.'
Said the
mouse to
the cur,
'Such a
trial, dear
sir. With
no jury
or judge,
would
be wast-
ting our
breath.'
'I'll be
judge,
I'll be
jury,'
said
cun-
ning
old
Fury:
'I'll
try
the
whole
cause,
and
con
demn
you to
death.'

SUSAN E. LEWAK

207

If the appearance of incoherence at the level of content conforms to the microlevel language of *no-sense*, it is the visual placement of the text which reveals the fact that Wonderland is indeed governed by a set of rules: as with the neatly aligned asterisks above, the meaning of "The Mouse's Tale," lies within the visual placement of text. Coherence, therefore, does exist at the microlevel, but it is multi-linear and visual, rather linear and text based.

Chapters VI-X outlines the second phase of Alice's journey, a period of slow acclimation and adjustment to the rules of Wonderland. Alice is now more interested in solving the word games presented to her than in re-establishing her former identity (or a static notion of scale). Indeed, one of the distinguishing marks of this section is Alice's growing ability to resolve some of Carroll's more intelligent puns, another indication of hidden sense behind the seemingly chaotic.

In order to master the rules of Wonderland, Alice must replace her macroworld notion of sense with the microworld vision of *no-sense*. Her anthropomorphizing of the Footman at the opening of Chapter VI indicates her initial difficulties with this transition: "she considered [him] to be a footman because he was in a livery; otherwise, judging by his face only, she would have called him a fish" (45). Indeed, Alice's hesitation to break from macroworld decorum left her temporarily unable to enter the Duchess' house: "How am I to get in?" (46) Alice asks repeatedly, only to receive a microworld response: " 'Are you to get in at all?' said the Footman. 'That's the first question, you know'" (46). It is at this juncture that Alice begins to realize why she can no longer recite poetry correctly. In Wonderland, one must accept the fact that the macroworld notion of madness is instead the microworld vision of sense: "'Oh there's no use talking to him,' said Alice desperately: 'he's perfectly idiotic'" (46).

This is further reinforced through Alice's meeting with the Cheshire Cat. As with the Footman, Alice initially clings to macroworld rules: "The Cat only grinned when it saw Alice. It looked good-natured, she thought" (51). However, just as Alice's attempts to follow macroworld decorum with the Footman proved futile, there is little correlation between the Cat's grin and its sense of integrity. The smile (a metaphor for the rules of Wonderland) could mean many things or nothing at all. Indeed, it is representative of what the Cat defines as microworld *no-sense*: "We're all mad here. I'm mad. You're mad . . . you must be or you wouldn't have come here" (51). Indeed, the Cat explains to Alice, this "madness" is the essence of Wonderland: "I growl when I'm pleased and wag my tail when I'm angry. Therefore I'm mad" (51).

Alice continues to grapple with the language of *no-sense* in one of the novel's most famous scenes, the "Mad Tea Party." When offered wine despite the absence of a wine bottle, Alice denounces the offer: "It wasn't very civil of you to offer it" (54). After exchanging a few nonsensical comments with the Mad Hatter, "Alice felt dreadfully puzzled. The Hatter's remark seemed to her to have no meaning in it, and yet it was certainly English. 'I don't understand you,' she said as politely as she could" (56). The comment does not elicit a response implying that as long as Alice ignores the rules of no-sense, Wonderland will appear chaotic.

It is when she meets with the Queen of Hearts that Alice begins to perceive order, rather than wonder at the microscale. She realizes, to her surprise that " they're only a pack of cards, after all. I needn't be afraid of them" (63). Thus, when Alice discovers that the Queen will threaten to behead anyone whenever she chooses, Alice negates this possibility by declaring: "Nonsense!" (64). In naming the language of Wonderland, Alice indicates her full assimilation into its realm: a place not of chaos but rather of an alternate form of order.

Alice's increasing fluency in the microworld language of *no-sense* is further indicated through the threatened "execution" of the Cheshire Cat who reappears after a frustrating game of flamingo croquet. Alice (having assimilated to microscale rules), politely waits for all of the body parts of the Cat to appear before she begins speaking to it as " 'it's no use speaking to it . . .'till its ears have come or at least one of them.' In another minute the whole head appeared . . .The Cat seemed to think that there was enough of it in sight, and no more of it appeared" (67). When the Queen threatens to behead the Cat for refusing to kiss her hand, Alice uses her growing fluency in the language of Wonderland to rescue it, arguing that before a beheading could take place, the owner of the Cat should be contacted. This new logic satisfies the queen despite the quiet disappearance of the head itself. As with the macroworld, upholding Wonderland decorum is of far greater importance than the actual execution itself.

Alice's assimilation into the *no-sense* language appears complete through her discussion of the dance, the "Lobster-Quadrille" with the Mock Turtle and the Gryphon. She is perplexed neither by the idea that lobsters and whiting are a normal part of the routine nor the Mock Turtle's assertion that: "we can do [the dance] without the lobsters" (79). Furthermore, her description of whiting as a dinner course ["tails in their mouths–and they're all over crumbs" (80)] is (mis) understood by the mock turtle who thinks that Alice merely misidentified them. The potential break in decorum (revelation of which would appear to the Mock Turtle as cannibalism) is thus overlooked in favor of Alice's new found ability to converse fluently in the language of *no-sense*.

Alice's assimilation into the microworld of Wonderland also carries a new realization. She is a different person who now lacks macroworld abilities: "it's no use going back to yesterday because I was a different person then" (81). Rather than confidently affirm that she can recite poetry in Wonderland, Alice now determines that it is a sheer impossibility: "her head was full of the Lobster-Quadrille, that she hardly knew what she was saying" (82). Thus, instead of hoping, as she did with the Caterpillar, that reverting to a macroworld size would restore her former sense of self, Alice "sat down with her face in her hands, wondering if anything would ever happen in a natural way again" (83).

In fact as Alice will soon discover (Chapters XI-XII) her return to the "natural" way of the macroworld is imminent. Alice is invited to the trial of the Knave, accused of stealing Tarts from the Queen. After arriving, Alice quietly listens to a trial she might have once deemed absurd. Indeed, her silence is an indication not only of a new-found fluency in the *no-sense* language, but also in the gradual loss of wonder which comes with understanding. This loss is most clearly evidenced through Alice's body, which resumes its macroworld state:

> Just at that moment Alice felt a very curious sensation, which puzzled her a good deal until she made out what it was: she was beginning to grow larger again and she thought at first she would get up and leave the court; but on second thoughts she decide to remain where she was as long as there was room for her" (88).

The loss of wonder is further evidenced through the inhabitants of Wonderland who now reject her: "Rule Forty-two, All persons more than a mile high to leave the court" (93). Perhaps in the hope of maintaining wonder, Alice demands that she be allowed to stay despite the fact that: "she had grown so large in the last few minutes that she wasn't a bit afraid of interrupting [the King]" (95). This desire, however, is overpowered by the realization that wonder itself is an illusion, one which no longer exists:

> "Who cares for you?" said Alice (she had grown to her full size by this time). "You're noth-
> ing but a pack of cards!" At this the whole pack rose up into the air and came flying down
> upon her; she gave a little scream, half of fright and half of anger, and tried to beat them
> off, and found herself lying on the bank, with her head in the lap of her sister (98).

Alice's departure from Wonderland, and thus from wonder itself, leads to a reunion with her sister (or the macroscale). While Alice does convey the experience of wonder to her sister (who briefly journeys to Wonderland herself through a further dream), neither return to it again, except perhaps through future recountings of the tale to their children. But a retelling is itself a type of macroworld logic: it is a means of giving order and meaning to the appearance of chaos, thus removing the possibility of wonder. Fully versed within the rules of Wonderland, Alice has lost the illusion of wonder she had associated with it. She thus attempts to find wonder in a different fantastic world (her adventures through the looking glass). Thus, it t is the sense of wonder, rather than the object that she associates with it, which she seeks.

Though Alice's attraction to the land of the White Rabbit lies in its potential for wonder, the value of either wonder or Wonderland is not necessarily diminished. As N. Katherine Hayles indicates in the intro-duction to this text, "If the cultural perception turns largely negative, this could have a damaging effect on funding sources and consequently on the field's development; conversely, if the public responds to its utopian possibilities and positive transformative social impact, this would also be important."[25] Wonder (i.e. the mode of cultural perception), is necessary to develop the kind of interest necessary in advancing the potential of nanotechnology. Likewise, despite the fact that "the buzz" surrounding nanotechnology will eventually diminish when some of the predicated "nano" changes are not met, what is realized in the future world of nanotechnology will vastly impact the way in which we live.

Visitors to the LACAMALab's Nano exhibit may make a similar discovery. If the journey through the museum space (propelled by the promise of wonder) is initially disorienting, the gradual familiariza-tion of the unfamiliar is reinforced through the final stop in the "Resource Room" (library) which explains the experience of the exhibit. Just as Alice grows to her normal size before she leaves Wonderland (thus conflating the macro with the microscale), the visit to the "Resource Room," merges macroscale logic with the experiences of the nano exhibit. The appearance of wonder may thus be deflated by the presence of a new set of rules, forcing visitors to realize (as Alice did) that the nano world is "nothing but a pack of cards!" (97). This familiarization, however, should not diminish the future role of nanotechnology in their lives, but instead should allow them to accept these changes as a natural phenomenon. By placing an experience of nanotechnology within the context of a museum space (often thought of as the place for art) the exhibit simultaneously mimics our ambivalence towards wonder as well as art's ability to evoke it.

○○○

Endnotes

Chapter 1

1 For a description of utility fog, see J. Storrs Hall, "Utility Fog: The Stuff That Dreams Are Made Of," in *Nantoechnology: Molecular Speculations on Global Abundance,* ed. B. C. Crandall (Cambridge: MIT Press, 1996), 161-184. On medical applications, see A. Paul Alivisatos, "Less Is More in Medicine," in *Understanding Nanotechnology: From the Editors of Scientific American,* comp. Sandy Fritz (New York: Warner Books, 2002), 56-71. On material abundance, see K. Eric Drexler, *Engines of Creation: The Coming Era of Nanotechnology* (New York: Anchor Books, 1986), 53-63 and passim.

2 This research is described in Michael L. Roukes, "Plenty of Room, Indeed," in *Understanding Nanotechnology,* 18-35.

3 George M. Whitesides and J. Christopher Love, "The Art of Building Small," in Understanding Nanotechnology, 326-55, especially p. 52; for quantum dots, see Tapash Chakraborty, *Quantum Dots* (New York: Elsevier Health Sciences, 1999).

4 Charles Ostman "The Creative and Design Intelligence Evolution in Today's Technological Innovation Roundtable," (presentation at "Visionary Forum 2003: Nanotechnology and Beyond," Jet Propulsion Laboratory, La Cañada Flintridge, CA, July 22, 2003). In an email dated October 29, 2003, he explains, "The reasons for this are actually surprisingly simple, [arising from] the precision and consistency with which nanoparticles can be created on a mass scale, [which] is particularly relevant to products where colloidal suspendants and related materials mixtures are critical to the performance of the product."

5 Richard Feynman, "There's Plenty of Room at the Bottom" (speech given at the Annual Meeting of the American Physical Society, California Institute of Technology, Pasadena, CA, December 29, 1959), available online at <http://www.zyvex.com/nanotech/feynman.html> (accessed October 20, 2003), originally published in *Engineering and Science* 23, no. 5 (February 1960): 22-36.

6 K. Eric Drexler, *Nanosystems: Molecular Machinery, Manufacturing, and Computation* (New York: John Wiley & Sons, 1992); and K. Eric Drexler, Chris Peterson, and Gayle Pergamit, *Unbounding the Future: The Nanotechnology Revolution* (New York: Quill, 1992).

7 James Gimzewski and Victoria Vesna, "The Nanomeme Syndrome: Blurring of Fact and Fiction in the Construction of a New Science," *Technoetic Arts Journal* 1, no. 1 (May 2003), 7-24, available online at <http://vv.arts.ucla.edu/publications/publications_frameset.htm> (accessed October 20, 2003).

8 R. Dean Astumian, "Making Molecules into Motors," in *Understanding Nanotechnology,* 72-86. See also Mark Ratner and Daniel Ratner, "Molecular Motors," in *Nanotechnology: A Gentle Introduction to the Next Big Idea* (Upper Saddle River, NJ: Prentice Hall, 2003), 114-115.

9 Drexler, *Engines of Creation,* 5.

10 The literature on the role of metaphor in science is extensive. For a sampling, see Laura Otis, *Membranes: Metaphors of Invasion in Nineteenth-Century Literature, Science and Politics* (Baltimore: Johns Hopkins University Press, 2000); and Ken Baake, *Metaphor and Science: The Challenge of Writing Science,* SUNY Series: Studies in Scientific and Technical Communication (New York: State University of New York Press, 2003).

11 *Understanding Nanotechnology,* 103.

12 *Understanding Nanotechnology,* viii.

13 Steven Ashley, "Nanobot Construction Crews," in *Understanding Nanotechnology,* 86-91.

14 For an announcement of Merkle's appointment, see the press release from the Georgia Tech College of Computing, <http://www.cc.gatech.edu/news/merkle.html> (accessed October 20, 2003).

15 Zyvex website, <http://zyvex.com/Products/home.html> (accessed October 20, 2003).

16 Ashley, "Nanobot Construction Crews," 980.

17 NanoInvestor News website, <http://www.nanoinvestornews.com> (accessed October 20, 2003).

18 Ostman, "The Creative and Design Intelligence Evolution."

19 Michael Crichton, *Prey: A Novel* (New York: Harper Collins, 2002).

20 Greg Bear, *Blood Music* (New York: I Books, 2002).

21 Neal Stephenson, *The Diamond Age: or, A Young Lady's Illustrated Primer* (New York: Spectra, 2000).

22 David E. Nye, *American Technological Sublime* (Cambridge: MIT Press, 1994).

23 Eric Drexler, *Engines of Creation,* 173.

24 Bill Joy, "Why the Future Doesn't Need Us Anymore," *Wired* 8, no. 4 (April 2000), <http://www.wired.com/wired/archive/8.04/joy.html> (accessed October 20, 2003).

25 Hans Moravec, *Mind Children: The Future of Robot and Human Intelligence* (Cambridge: Harvard University Press, 1988).

26 Vernon Vinge, "The Coming Technological Singularity: How to Survive in the Post-Human Era" (VISION-21 Symposium, NASA Lewis Research Center and the Ohio Aerospace Institute, March 30-31, 1993), <http://www.cse.ucsd.edu/users/goguen/misc/singularity.html> (accessed October 20, 2003).

27 Neils Bohr took over this term (its usual meaning is the numerical value of Planck's constant) to denote an area within which no further distinction between the observer and system is possible.

28 This favorite saying of Bohr is documented by Aage Petersen, "The Philosophy of Neils Bohr," *Bulletin of the Atomic Scientist* 19 (September 1963), 10.

Chapter 2

1 I would like to thank CAPES Foundation (Coordination for the Development of Graduate and Academic Research, Brazil) and CNPq (National Research Council, Brazil) supports for providing Doctoral fellowships. Thanks to Ashok Sukumaran for the exhibition drawings and to Andrew Pelling and James Gimzewski for the information about nanoscience. Many thanks to Daniel Sauter for all his help and support.

2 All images in this essay are illustrations developed during the process of producing the exhibition. While they are not photo-realistic pictures from the show, they represent the early intentions of artists and scientists involved in its development. Therefore they can be viewed as potential spaces that will be actualized in the final *nano* show. The final pictures can be found in a special signature in the middle of the book.

3 This is a fictional walk-through the *nano* exhibition by an imaginary visitor. This description does not intend to cover the entire show, for its richness and complexity exceeds the scope of this essay. Elements that belong to the exhibition but have not been mentioned: *Seeing by Feeling,* the *Mandala Kaleidoscope, Virtual Crystal Assembly, Model Assembly,* and the *Lounge Area with*

Nanofurniture. This was the status of the exhibition in December 2003. As the exhibit continues to evolve, some of its components may also change.

4   For a more detailed discussion about nanotechnology and surveillance mechanisms, please refer to the chapter, "Less is More: Much Less Much More: the Insistent Allure of Nanotechnology Narratives in Science Fiction," by Brooks Landon, *infra* 131-146

5   The exceptions are viruses.

6   The installation is based on *Zero@wavefunction: nanodreams and nightmares,* the first collaboration between artist Victoria Vesna and scientist James Gimzewski. <http://notime.arts.ucla.edu/zerowave>.

7   Robert Sain, LACMALab Overview Catalogue.

8   Foucault, Michel. *Of Other Spaces* (1967), *Heterotopias.* <http://foucault.info/documents/foucault.heteroTopia.en.html> (October 16, 2003).

9   Manovich, Lev. *The Poetics of Augmented Spaces: Learning from Prada.* <http://www.manovich.net/> p. 11. (May 25,2003).

10 Manovich, Lev. *The Poetics of Augmented Spaces: Learning from Prada.* <http://www.manovich.net/> p. 12. (May 25,2003).

11 Stalder, Felix. "The Space of Flows: notes on emergence, characteristics and possible impact on physical space." <http://felix.openflows.org/html/space_of_flows.html> (May 30, 2003).

12 For information, see <http://www.connected-cities.de/>.

13 For a more detailed explanation of Eigler and Schweizer's experience, please refer to Nathan Brown's article in this book, "Needle on the Real: Technoscience and Poetry at the Limits of Fabrication," *infra* 173-190.

14 Gimzewski, James (interviewed by Rebecca N. Lawrence). *Magazine: BioMedNet.* <http://news.bmn.com/hmsbeagle/118/notes/biofeed> (May 30, 2003).

15 Gimzewski, James (interviewed by Rebecca N. Lawrence). *Magazine: BioMedNet.* <http://news.bmn.com/hmsbeagle/118/notes/biofeed> (May 30, 2003).

16 For more dreams and nightmares, visit *Zero@wavefunction* website at <http://notime.arts.ucla.edu/zerowave>.

17 Weiser, Mark and Brown, John Seely. *Designing Calm Technology.* <http://www.ubiq.com/weiser/calmtech/calmtech.htm> (August 25, 2003).

18 William Blake, *The Complete Writings of William Blake,* edited by Geoffrey Keynes (London: Oxford University Press, 1966), letter to Dr. Trusler, 23 August 1799, p. 793. The capitalization is Blake's.

Chapter 3

1   Thanks to everyone who took the time to read and comment on drafts of this essay: Katherine Hayles, Victoria Vesna, James Gimzewski, Robert Sain, Carol Eliel, Sharon Johnston, Nathan Brown, Susan Lewak, Jessica Pressman, Adriana de Souza e Silva, Andrew Pelling, Nicholas Gessler, Patrick Sharp, Sharon Sharp, Margaret Luesebrink, and Tony Jackson.

2   Julie Thompson Klein, *Crossing Boundaries: Knowledge, Disciplinarities, and Interdisciplinarities* (Charlottesville: University Press of Virginia, 1996).

3   Since the names of all but one of the principal collaborators are well known and clearly identified in the exhibition, there is no reason to use pseudonyms, except in the case of Williamson, the science writer. Williamson resigned from the project in June 2003. Ultimately, she declined to be interviewed for this essay, and asked to remain anonymous. To represent her point of view, I relied

on informal conversations between Williamson and myself, as well as Hayles' and Sain's accounts of their conversations with Williamson. Students agreed to be interviewed with the condition of anonymity, and so are not named.

4   *Crossing Boundaries*, 1. Klein attributes the concept of boundary work to the research of sociologists of science Thomas Giernyn and Steven Shapin.

5   The "synaptic blow-out" emerged from a collaborative project begun in 2001 by Hayles and Vesna called SINAPSE (Center for Social Interfaces and Networks, Advanced Programmable Simulations and Environments), which Gimzewski later joined. SINAPSE was conceived as a "catalyst to bring together researchers in digital media for interdisciplinary conversations and collaborations, coordinate the activities of Centers and Institutes concerned with digital media so that they perform not only as individual research units but also as seedbeds for interdisciplinary collaborative work, provide the environment, direction, and support for the development of grant proposals in digital media, and study how IT is used to foster a interdisciplinary and collaborative environment" *(<http://sinapse.arts.ucla.edu/background.html>)*. SINAPSE sponsored the UCLA panels at the 2001 conference "Networks to Nanosystems", a UC-wide conference sponsored by UC DARNET (University of California Digital Arts Research Network); Vesna was a co-organizer of the UCLA panels.

6   *Crossing Boundaries*, 1.

7   Ibid.

8   <http://notime.arts.ucla.edu/zerowave/>

9   Crista Sommerer and Laurent Mignonneau, eds., *Art @ Science* (New York: SpringerWeinNewYork, 1998), 3, 13.

10 The 2001 lecture series at which Vesna and Hayles met was a regular Monday night event, organized by Vesna, featuring talks by visiting media artists.

11 Ibid.

12 The source of this and all other Hayles quotations is an interview I conducted on 7/13/03.

13 Lecture at UCLA, 4/4/03.

14 "Point of View: Transdisciplinarity on a Solid Basis," Annual Report, Swiss Research Council, 1998, <http://www.snf.ch/en/por/phi/phi_view_hd1.asp>. See also *Transdisciplinarity: Joint Problem-Solving among Science, Technology, and Society,* eds. Julie Thompson Klein, et. al. (Basel: Birkhäuser Verlag, 2001); *Transdisciplinarity: Recreating Integrated Knowledge,* David Rapport and Margaret A. Somerville (Montréal: McGill-Queen's University Press), 2003.

15 The source of this and all subsequent quotations from Sain is an interview I conducted on 9/29/03.

16 The source of this and all subsequent quotations from Eliel is an interview conducted on 11/20/03.

17 E-mail communication, 12/4/04.

18 Interview with James Gimzewski, 10/3/03.

19 Ibid.

20 Joint interview with Gimzewski and Vesna, 8/22/03.

21 Joint interview with Vesna and Gimzewski, 8/22/03.

22 Vesna asserted later that she had never viewed Williamson as an equal collaborator (e-mail communication, 11/21/03); but Sain subsequently confirmed to me that he believed he had made this clear to all of the principals when he invited Williamson to participate, and that "everyone was to be equal on the team" (e-mail communication, 11/24/03).

23 During interviews, Sain and Hayles both expressed their sense of being sidelined in the conceptualization phase of the project. In informal conversation, Williamson expressed essentially the same perception.

24 E-mail communication, 11/23/03.

25 The disagreement over the use of excerpts from *Prey* and *Engines of Creation* points to a deeper difference between the disciplinary investments of Vesna and Gimzewski on the one hand, and Hayles on the other. Hayles told me she believed it was made very clear by Sain at the outset of the project that the goal of the exhibition was not didactic, but rather to explore various representations of nanotechnology. Gimzewski, on the other hand, did see the exhibition as having a didactic function. In my interview with him on 10/3/03, he talked about his conviction that exposing young people to sensory, non-verbal representations of key nanoscience concepts could constitute a kind of "unconscious education" that would prepare children's minds to encounter abstract versions of these concepts in school. Later, he asserted the didactic purpose for the exhibition more bluntly: "The whole point of our [exhibition] concept is to stop propagating the ideas that 'have had an impact on popular beliefs about nanotechnology.' The U.S. and U.K. government[s] and 99% of technology companies share this objection" (e-mail communication, 11/21/03). For a full articulation of Vesna and Gimzewski's argument on this point, see James Gimzewski and Victoria Vesna, "The Nanomeme Syndrome: Blurring of fact and fiction in the construction of a new science," *Technoetic Arts,* v. 1, n. 1, 7-24.

26 My account of the exchange between Hayles and Vesna at this meeting is based on written notes; all other quotations throughout this essay are based on recorded interviews.

27 This and following Vesna quotations, e-mail communication, 11/23/03.

28 For an expansion upon Vesna's argument regarding, holism, the separation between theory and practice in the academy, and its relation to C.P. Snow's "two cultures" problem, see Victoria Vesna, "Toward a Third Culture: Being in Between," *Leonardo,* v. 34, n. 2 (April 2001). Available at <http://vv.arts.ucla.edu/publications/publications_frameset.html>.

29 For a more detailed analysis of the role of holism in transdisciplinarity, and transdisciplinarity's ideological potential, see *Crossing Boundaries,* 13-15.

30 Hayles' view that writing practices are material embodiments represents a major motif in her more recent work, such as *Writing Machines* (Cambridge: The MIT Press, 2002). See also Hayles, "The Materiality of Informatics," in *How We Became Posthuman: Virtual Bodies and Flickering Signifiers* (Chicago: University of Chicago Press, 1999); Tim Lenoir, ed., *Inscribing Science: Scientific Texts and the Materiality of Communication* (Stanford: Stanford University Press, 1998).

31 Vesna introduced Hayles to the publisher of this book, with the idea that the journal *Technoetic Arts,* also published by Intellect Books, might want to run a special issue on the exhibition, guest-edited by Hayles. In subsequent talks with the publisher, Hayles proposed a book instead, and the contract was signed by Hayles. Hayles and Vesna jointly submitted a grant application to fund the symposium.

32 E-mail communication from Vesna to myself, 11/21/03.

33 The source of this and following Gimzewski quotations is an interview I conducted on 10/3/03.

34 Student interview, 8/20/03.

35 E-mail communication, 11/24/03.

36 Joint interview with Vesna and Gimzewski, 8/22/03.

Chapter 4

My thanks to Mario Biagioli, Robert Brain, Timothy Lenoir, and an anonymous reader for their many wonderful suggestions and generous support throughout the development of this chapter. "Nanotechnology in the Age of Posthuman Engineering: Science Fiction as Science" has been previously published in *Configurations* 10:2 (2002), 261-296, © Johns Hopkins University Press, and I am grateful to them for permission to re-publish it here.

1 K. Eric Drexler, "Preface," in K. Eric Drexler, Chris Peterson, and Gayle Pergamit, *Unbounding the Future: The Nanotechnology Revolution* (New York: Morrow, 1991), p. 10.

2 For a complete catalogue of nanotechnology research sites established up through 2002, see "Nanotechnology Database" (2002), International Technology Research Institute, World Technology (WTEC) Division, <www.wtec.org/loyola/nano/links.htm>.

3 The National Nanofabrication Users Network, <www.nnun.org/>, "provides users with access to some of the most sophisticated nanofabrication technologies in the world with facilities open to all users from academia, government, and industry."

4 "National Nanotechnology Initiative" (2003), <www.nano.gov>.

5 The Foresight Institute, <www.foresight.org>, is based in Palo Alto, California.

6 K. Eric Drexler, *Engines of Creation: The Coming Era of Nanotechnology,* (Garden City, N.Y.: Anchor Books/Doubleday, 1986); all references to this work are to the revised ed. (New York: Anchor Books/Doubleday, 1990).

7 Chad A. Mirkin, "Tweezers for the Nanotool Kit," *Science* 286 (1999): 2095-2096, on p. 2095.

8 Robert F. Service, "AFMs Wield Parts for Nanoconstruction," *Science* 282 (1998): 1620-1621, on p. 1620.

9 Robert F. Service, "Borrowing From Biology to Power the Petite," *Science* 283 (1999): 27-28, on p. 27.

10 James K. Gimzewski and Christian Joachim, "Nanoscale Science of Single Molecules Using Local Probes," *Science* 283 (1999): 1683-1688, on p. 1683.

11 Richard Smalley, "Nanotech Growth," *Research and Development* 41.7 (1999): 34-37.

12 Christine L. Peterson, "Nanotechnology: Evolution of the Concept," in *Prospects in Nanotechnology: Toward Molecular Manufacturing,* eds. Markus Krummenacker and James Lewis (New York: Wiley, 1995), pp. 173-186, quotation on p. 186. Indicative of nanowriting's teleological tendencies, Peterson's article absorbs the entire history of atomic theory, from Democritus to the present, to suggest the unavoidable rise of nanotechnology and our progression toward the nanofuture.

13 Nanowriting employs literary techniques common to speculative science writing in general. See Greg Myers, "Scientific Speculation and Literary Style in a Molecular Genetics Article," *Science in Context* 4 (1991): 321-346, on the linguistic peculiarities of scientific speculation that work to legitimate such claims. Nanowriting, however, goes beyond most scientific speculation in that its uses of the future tense and its visions of tomorrow are totalizing, bringing the future more firmly into the textual present—which is one reason, as we will see, why nanotechnology is so frequently characterized not as "speculative science" but as "fictional science."

14 B.C. Crandall, "Preface," in *Nanotechnology: Molecular Speculations on Global Abundance,* ed. idem (Cambridge, Mass.: MIT Press, 1996), pp. ix-xi, quotation on p. ix.

15 K. Eric Drexler, "Molecular Engineering: An Approach to the Development of General Capabilities for Molecular Manipulation," *Proceedings of the National Academy of Sciences* 78 (1981): 5275-5278, on p. 5278.

16 David E. H. Jones, "Technical Boundless Optimism," *Nature* 374 (1995): 385-387, on pp. 835, 837.

17 Gary Stix, "Trends in Nanotechnology: Waiting for Breakthroughs," *Scientific American* 274.4 (1996): 94-99, on p. 97.

18 Gary Stix, "Little Big Science," *Scientific American* 285.3 (2001): 32-37, on p. 37.

19 Steven M. Block, "What is Nanotechnology?" keynote presentation at the National Institutes of Health conference, "Nanoscience and Nanotechnology: Shaping Biomedical Research," Natcher Conference Center, Bethesda, Maryland, June 25, 2000 (quotation from program abstract).

20 Many early critiques of nanotech's "science-fictionality" are described in Ed Regis's lively history, *Nano: The Emerging Science of Nanotechnology* (Boston: Little, Brown, 1995).

21 Darko Suvin, *Metamorphoses of Science Fiction: On the Poetics and History of a Literary Genre* (New Haven: Yale University Press, 1979), p. 64.

22 Ibid., p. 75.

23 The fabulation of worlds and zones of radical otherness characterizes the interplay between science fiction and postmodernist writing; see Brian McHale, *Postmodernist Fiction* (New York: Routledge, 1997), esp. pp. 59-72.

24 Jean Baudrillard, *Symbolic Exchange and Death,* trans. Iain Hamilton Grant (London/ Thousand Oaks/New Delhi: Sage Publications, 1993), pp. 50-86. See also Baudrillard, "The Precession of Simulacra," in idem, *Simulacra and Simulation,* trans. Sheila Faria Glaser (Ann Arbor: University of Michigan Press, 1994), pp. 1-42.

25 Baudrillard, "Simulacra and Science Fiction," in idem, *Simulacra and Simulation,* pp. 121-127, quotation on p. 121.

26 Ibid.

27 Ibid., p. 122 (emphasis in original).

28 Ibid., p. 124.

29 Ibid., pp. 125, 126.

30 Donna J. Haraway, "A Cyborg Manifesto: Science, Technology, and Socialist-Feminism in the Late Twentieth Century," *in idem, Simians, Cyborgs, and Women: The Reinvention of Nature* (New York: Routledge, 1991), pp. 149-181, quotation on p. 149.

31 Scott Bukatman, *Terminal Identity: The Virtual Subject in Postmodern Science Fiction* (Durham, N.C.: Duke University Press, 1993), p. 22.

32 N. Katherine Hayles, *How We Became Posthuman: Virtual Bodies in Cybernetics, Literature, and Informatics* (Chicago: University of Chicago Press, 1999), p. 3.

33 Ralph C. Merkle, "It's a Small, Small, Small, Small World," *Technology Review* 100.2 (1997): 25-32, on p. 26 (emphasis added).

34 These science-fictional images and a general faith in the imminent nanofuture can be found in, but not limited to: Drexler, "Molecular Engineering" (above, n. 15), pp. 5275-5278; Drexler, "Molecular Manufacturing as a Path to Space," in Krummenacker and Lewis, *Prospects in Nanotechnology,* pp. 197-205; Ralph C. Merkle, "Nanotechnology and Medicine," in *Advances in Anti-Aging Medicine,* vol. 1, eds. Ronald M. Klatz and Francis A. Kovarik (Larchmont, NY: Liebert, 1996), pp. 277-286; Robert A. Freitas, Jr., *Nanomedicine, Volume I: Basic Capabilities* (Georgetown, Tex: Landes Bioscience, 1999); Markus Krummenacker, "Steps Towards Molecular Manufacturing," *Chemical Design Automation News* 9 (1994): 1, 29-39; Daniel T. Colbert and Richard E. Smalley, "Fullerine Nanotubes for Molecular Electronics," *Trends in Biotechnology* 17 (1999): 46-50; J. Storrs Hall, "Utility Fog: The Stuff that Dreams Are Made Of," in Crandall, *Nanotechnology* (above, n. 14), pp. 161-184.

35 This argument is ubiquitous in nanowriting: Drexler, *Engines of Creation* (above, n. 6), pp. 5-11; Drexler, "Molecular Engineering" (above, n. 15), pp. 5575-5576; Merkle, "Molecular Manufacturing: Adding Positional Control to Chemical Synthesis," *Chemical Design Automation News* 8 (1993): 1, 55-61; Merkle, "Self-Replicating Systems and Molecular Manufacturing," *Journal of the British Interplanetary Society* 45 (1992): 407-413; Service, "Borrowing from Biology to Power the Petite" (above, n. 9), p. 27.

36 J.S. Foster, J.E. Frommer, and P.C. Arnett, "Molecular Manipulation Using a Tunneling Microscope," *Nature* 331 (1988): 324-326.

37 D.M. Eigler and E.K. Schweizer, "Positioning Single Atoms with a Scanning Tunneling Microscope," *Nature* 344 (1990): 524-526.

38 Buckminsterfullerines, a.k.a. "buckyballs" or C_{60}, can potentially serve as robust structural or conductive materials, or as containers for individual atoms. Other fullerines called "nanotubes"—discovered by Sumio Iijima—can serve as probes or funnels for atomic positioning. See Hongjie Dai, Nathan Franklin, and Jie Han, "Exploiting the Properties of Carbon Nanotubes for Nanolithography," *Applied Physics Letters* 73 (1998): 1508-1510; Philip Kim and Charles M. Lieber, "Nanotube Nanotweezers," *Science* 286 (1999): 2148-2150. The hypothetical applications of fullerines are extraordinary—but remain to be realized in the future.

39 Anthony P. Davis, "Synthetic Molecular Motors," *Nature* 401 (1999): 120-121; M.T. Cuberes, RR. Schlittler and J.K. Gimzewski, "Room-Temperature Repositioning of Individual C60 Molecules at Cu Steps: Operation of a Molecular Counting Device," *Applied Physics Letters* 69 (1996): 3016-3018; Jonathan Knight, "The Engine of Creation," *New Scientist* 162.2191 (1999): 38-41.

40 On confrontations within nanotechnology to capture outside credibility, see David Rotman, "Will the Real Nanotech Please Stand Up?" *Technology Review* 102.2 (1999): 47-53. For analysis of the social dynamics of scientific discipline formation, see Timothy Lenoir, *Instituting Science: The Cultural Production of Scientific Disciplines* (Stanford: Stanford University Press, 1997).

41 Drexler, *Engines of Creation* (above, n. 6), p. 241.

42 Drexler, *Nanosystems: Molecular Machinery, Manufacturing, and Computation* (New York: Wiley, 1992).

43 Zyvex, <www.zyvex.com>.

44 Drexler, *Engines of Creation* (above, n. 6), pp. 92-93.

45 This strategy depends somewhat on a lack of definitional precision in the term "nanotechnology." Many researchers adopt the term for work in nanoelectronics, nanocomposition, or nanolithography that does not necessarily match Drexler's definition of nanotechnology as molecular manufacturing.

46 Eigler quoted in Rotman, "Will the Real Nanotech Please Stand Up?" (above, n. 40), p. 53.

47 Reed quoted in ibid., p. 48 (emphasis added).

48 Richard Feynman, "There's Plenty of Room at the Bottom," in *Miniaturization,* ed. H.D. Gilbert (New York: Reinhold, 1961), pp. 282-96, quotation on p. 295; all further references to Feynman's 1959 talk are to this transcript. The talk was originally published in *Engineering and Science* 23 (February 1960): 22-36; a shorter version was published as "The Wonders that Await a Micro-microscope," *Saturday Review* 43 (1960): 45-47; also available on the web at <www.zyvex.com/nanotech/feynman.html>. These and several other published incarnations of Feynman's talk, each with independent legacies of citation, suggest the prevalence and influence of the speech within the technoscape.

49 A few examples among many: Drexler, "Molecular Engineering" (above, n. 15), p. 5275; Drexler, *Engines of Creation* (above, n. 6), pp. 40-41; Davis, "Synthetic Molecular Motors" (above, n. 39), p. 120; Gimzewski and Joachim, "Nanoscale Science" (above, n. 10), p. 1683; Merkle, "Nanotechnology and Medicine" (above, n. 34), pp. 277-86; Merkle, "Letter to the Editor," *Technology Review* 102.3 (1999): 15-16. See also Regis, *Nano* (above, n. 20), pp. 10-12, 63-94, for a history of this talk and its sociological function.

50 J. Storrs Hall, "An Overview of Nanotechnology" (1995), <nanotech.dyndns.org/sci.nanotech/overview.html>.

51 "Richard Feynman" (2002), *Photosynthesis.Sound.com*, <www.photosynthesis.com/Richard_Feynman.html>.

52 Merkle, "A Response to *Scientific American*'s news story *Trends in Nanotechnology*" (1996), <www.islandone.org/Foresight/SciAmDebate/SciAmResponse.html>.

53 The nanotechnologists' 1992 testimonies are found in *New Technologies for a Sustainable World: Hearing Before the Subcommittee on Commerce, Science and Transportation, United States Senate, One Hundred Second Congress, Second Session, June 26, 1992* (Washington, D.C.: U.S. Government Printing Office, 1993). The 1999 testimonies are found in *Nanotechnology: The State of Nanoscience and Its Prospects for the Next Decade: Hearing Before the Subcommittee on Science, House of Representatives, One Hundred Sixth Congress, First Session, June 22, 1999* (Washington, D.C.: U.S. Government Printing Office, 2000). Federal documentation for the National Nanotechnology Initiative is compiled by the National Science and Technology Council in *National Nanotechnology Initiative: Leading the Next Industrial Revolution* (Washington, D.C.: Office of Science and Technology Policy, 2000).

54 William J. Clinton, [Address to Caltech on Science and Technology], California Institute of Technology, January 21, 2000. Video transcript of talk produced by Caltech's Audio Visual Services, Electronic Media Publications, and Digital Media Center; made available online at <pr.caltech.edu/events/presidential_speech//PresVisit-MCP-LAN.ram>.

55 K. Eric Drexler, "From Nanodreams to Realities," introduction to *Nanodreams*, ed. Elton Elliott (New York: Baen Books, 1995), pp. 13-16. Drexler's argument is similar to that of Arthur C. Clarke's *Profiles of the Future: An Inquiry into the Limits of the Possible* (New York: Harper & Row, 1958). Frequently cited in Drexler's publications, Clarke's text aggressively foregrounds the science-fictional foundations of scientific extrapolation.

56 Drexler, *Engines of Creation* (above, n. 6), pp. 234-235.

57 Baudrillard, "The Precession of Simulacra" (above, n. 24), pp. 1-42.

58 Not the least of which are those technology companies founded in the 1990s to pursue some aspect of Drexler's vision, such as Zyvex, Nanogen, Molecular Manufacturing Enterprises, NanoTechnology Development Corporation, and NanoLogic.

59 Smalley quoted in David Voss, "Moses of the Nanoworld," *Technology Review* 102.2 (1999): 60-62, on p. 62. Smalley nevertheless states that Drexlerian self-replicating assemblers are "not possible" and believes nanotech will develop along different lines; see Smalley, "Of Chemistry, Love and Nanobots," *Scientific American* 285.3 (2001): 76-77.

60 See Regis, *Nano* (above, n. 20), pp. 3-18.

61 Merkle, "Letter to the Editor" (above, n. 49), p. 15.

62 Jonathan Culler, *The Pursuit of Signs: Semiotics, Literature, Deconstruction* (London: Routledge & Kegan Paul, 1981), p. 38.

63 Neal Stephenson, *The Diamond Age* (New York: Bantam Books, 1995); Merkle, "It's a Small, Small, Small, Small World" (above, n. 33), pp. 27, 29-31.

64 Stephenson, *The Diamond Age,* pp. 41-42.

65 Hall, "Utility Fog" (above, n. 34), pp. 161-84.

66 Thomas N. Thetis, "Letter to the Editor," *Technology Review* 102.2 (1999): 15.

67 See Regis, *Nano* (above, n. 20), pp. 152-154.

68 Robert A. Heinlein, "Waldo" (1942), in idem, *Waldo & Magic, Inc.* (New York: Dell Rey, Ballantine Books, 1986), pp. 1-154, quotations on pp. 29, 133.

69 Feynman, "There's Plenty of Room" (above, n. 48), p. 292.

70 Heinlein, "Waldo" (above, n. 68), p. 133.

71 Jacques Derrida, "Structure, Sign and Play in the Discourse of the Human Sciences," in idem, *Writing and Difference,* trans. Alan Bass (Chicago: University of Chicago Press, 1978), pp. 278-293, quotation on p. 292.

72 Jacques Derrida, "The Ends of Man," in idem, *Margins of Philosophy,* trans. Alan Bass (Chicago: University of Chicago Press, 1982), pp. 109-136, quotation on p. 123.

73 Michel Foucault, *The Order of Things: An Archaeology of the Human Sciences* (New York: Vintage Books, 1973), p. 387. It is worth noting that Derrida, "without naming names," critiques Foucault's claim for the end of man as a reinscription of humanist eschatology. Although ultimately affirming the necessity of Foucault's move outside the boundaries of humanism, Derrida suggests that this strategy must be accompanied by deconstruction from within in order to become other than another sedimented humanism (See Derrida, "The Ends of Man," pp. 114-123, 134-136).

74 See Paul Virilio, *The Art of the Motor,* trans. Julie Rose (Minneapolis: University of Minnesota Press, 1995); Brian Massumi, *Parables for the Virtual: Movement, Affect, Sensation* (Durham, N.C.: Duke University Press, 2002); Hayles, *How We Became Posthuman* (above, n. 32); and Hayles, "Flesh and Metal: Reconfiguring the Mindbody in Virtual Environments," *Configurations* 10 (2002): 297-320.

75 Kelly Hurley, "Reading Like an Alien: Posthuman Identity in Ridley Scott's Alien and David Cronenberg's Rabid," in *Posthuman Bodies, eds. Judith Halberstam and Ira Livingston* (Bloomington: Indiana UP, 1995), pp. 203-224, quotations on pp. 205, 220, 220, 205.

76 This alignment is strengthened by the commitment of several nanoscientists to the ideology of extropianism, which maintains that human life is evolving through the mediation of technosciences and the active pursuit of science fiction scenarios. 0Max More, president of the Extropy Institute, writes that extropians "advocate using science to accelerate our move from human to a transhuman or posthuman condition," and that nanotechnology is a significant vehicle in bringing about the posthuman future; see Max More, "The Extropian Principles 3.0" (1998), <www.extropy.com/ideas/principles.html>. Merkle and Drexler have lectured at several extropian conferences and Hall is nanotechnology editor of *Extropy: The Journal of Transhumanist Solutions.*

77 Hayles, *How We Became Posthuman* (above, n. 32), esp. pp. 1-24.

78 Drexler, *Engines of Creation* (above, n. 6), p. 38.

79 Feynman, "There's Plenty of Room" (above, n. 48), p. 295.

80 Merkle, who left Xerox in 1999 to become Principle Fellow of Zyvex, jokes about copy culture's impact on nanotech in "Design-Ahead for Nanotechnology," in Krummenacker and Lewis, *Prospects in Nanotechnology* (above, n. 12), pp. 23-52, on p. 35.

81 Hurley, "Reading Like a Alien" (above, n. 75), p. 205.

82 Drexler, *Engines of Creation* (above, n. 6), p. 103.

83 The fantasy of "telegraphing a human" appears frequently within posthuman science (fiction): for example, Norbert Wiener, *The Human Use of Human Beings: Cybernetics and Society* (Boston: Houghton Mifflin, 1954), pp. 95-104, and Hans Moravec, *Mind Children: The Future of Robot and Human Intelligence* (Harvard: Harvard University Press, 1988), pp. 116-122.

84 The "grey goo" hypothesis imagines the entire organic world dismantled into disorganized material: Drexler, *Engines of Creation,* pp. 172-173; Regis, *Nano* (above, n. 20), pp. 121-124. This horrifying possibility has so disturbed Bill Joy, cofounder and Chief Scientist of Sun Microsystems (and certainly no technophobe), that he questions the wisdom of pursuing nanoresearch, suggesting the future would perhaps be a better place if we did not follow our nanodreams. See Bill Joy, "Why the Future Doesn't Need Us," *Wired* 8.4 (2000): 238-263.

85 For a sustained discussion of nanotechnology's significance for and commitment to cryonics, see Wesley M. Du Charme, *Becoming Immortal: Nanotechnology, You, and the Demise of Death* (Evergreen, Colo.: Blue Creek Ventures, 1995). In Du Charme's account—typical of nanowriting— "you" are already a posthuman subject of the future.

86 The Alcor Foundation, founded in 1972, promotes public awareness of cryonic possibilities and assists its members in arranging for cryonic suspension after their deaths. Alcor cites fully functional nanotechnology as the fundamental scientific development still needed in order to repair and resurrect suspended patients. See "Alcor Life Extension Foundation" (2003), <www.alcor.org>. The Extropy Institute, in which many nanoscientists are involved (n. 76, above), traces its historical origins to the Alcor Foundation, signaling the deeply complicit nature of posthuman discourse and cryo-nanotechnology. For more on the interlinkage of cryonics, nanotechnology, and posthuman immortality, see Ed Regis, *Great Mambo Chicken and the Transhuman Condition: Science Slightly Over the Edge* (New York: Addison-Wesley, 1990), pp. 1-9, 76-143. See also Merkle's cryonics website, "Cryonics" (2003), <www.merkle.com/cryo/>.

87 Drexler, "Molecular Engineering" (above, n, 15), p. 5278.

88 Drexler, *Engines of Creation,* pp. 136-138.

89 Jean Baudrillard, *Écran Total* (Paris: Éditions Galilée, 1997), pp. 169-173; Baudrillard, "The Precession of Simulacra" (above, n. 24), pp. 12-14.

90 See Bukatman, *Terminal Identity* (above, n. 31), pp. 227-229; Bukatman, "There's Always Tomorrowland: Disney and the Hypercinematic Experience," *October* 57 (1991): 55-78.

91 Tom McKendree, "Nanotech Hobbies," in Crandall, *Nanotechnology* (above, n. 14), pp. 135-144, quotations on p. 143.

92 Merkle, "It's a Small, Small, Small, Small World" (above, n. 33), p. 26.

93 Regis, *Great Mambo Chicken and the Transhuman Condition* (above, n. 86), pp. 2, 126-30.

94 Merkle, "It's a Small, Small, Small, Small World," p. 32. Merkle actually cautions that nanotech will not develop on its own, but then strangely implies that even if we do not strive for the nanofuture, if "we ignore it, or simply hope that someone will stumble over it," the evolution of nanotech will still proceed—it is still inevitable—but it just "will take much longer."

95 Stephenson, *The Diamond Age* (above, n. 63), p. 31 (emphasis in original).

96 Donna J. Haraway, "The Biopolitics of Postmodern Bodies: Constitutions of Self in Immune System Discourse," in idem, *Simians, Cyborgs, and Women* (above, n. 30), pp. 203-230, quotation on p. 230. Chris Hables Gray persuasively reiterates this need for "participatory cyborg evolution" in his recent

political mapping of posthumanism, *Cyborg Citizen: Politics in the Posthuman Age* (New York: Routledge, 2001).

Chapter 5

1 Brian Stableford, "Great and Small," in *The Encyclopedia of Science Fiction,* ed. John Clute and Peter Nicholls (New York: St. Martin's Press, 1993), 519.

2 Thomas D. Clareson, *Some Kind of Paradise: The Emergence of American Science Fiction* (Westport, CT: Greenwood Press, 1985), 83-84.

3 John Huntington, *Rationalizing Genius: Ideological Strategies in the Classic American Science Fiction Short Story* (New Brunswick: Rutgers Univ Press, 1989), 54.

4 Anthony S. Napier has compiled the most extensive bibliography of nanotech narratives in SF and his list, which also indicates the degree to which a text is concerned with nanotech and in many cases its focus, contains over 280 titles: "The Nanotechnology in Science Fiction Bibliography," Issue 29, December 5, 2002. <http://www.geocities.com/asnapier/nano/n-sf/>

5 Wil McCarthy, "Nanotechnology: Abuses of, and Replacements For," *The Bulletin of the Science Fiction and Fantasy Writers of America* 151 (2001): 20.

6 Ibid., 21.

7 Ibid.

8 Graham P. Collins, "Shamans of Small," *Scientific American* 285, no. 3 (2001): 90.

9 Ibid.

10 John G. Cramer, "Nanotechnology: The Coming Storm," in *Nanodreams,* ed. Elton Elliott (New York: Baen Books, 1995), 10.

11 K. Eric Drexler, *Engines of Creation: The Coming Era of Nanotechnology* (New York: Anchor Books, 1986), 234.

12 K. Eric Drexler, "From Nanodreams to Reality," in *Nanodreams,* ed. Elton Elliott (New York: Baen Books, 1995), 15.

13 Elton Elliott, *Nanodreams* (New York: Baen Books, 1995), 2.

14 See N. Katherine Hayles, *How We Became Posthuman: Virtual Bodies in Cybernetics, Literature, and Informatics* (Chicago: University of Chicago Press, 1999).

15 Collins, "Shamans of Small," 86.

16 Tony Daniel, *Metaplanetary: A Novel of Interplanetary Civil War* (New York: EOS, 2001), 134.

17 Neal Stephenson, *The Diamond Age: Or, a Young Lady's Illustrated Primer* (New York: Bantam, 1995), 338.

18 Wil McCarthy, *Bloom* (New York: Del Rey Books, 1998), 43.

19 Octavia Butler, *Parable of the Sower* (New York: Four Walls Eight Windows, 1993), 3.

20 Kevin J. Anderson and Doug Beason, *Assemblers of Infinity* (New York: Bantam Spectra, 1993). 104.

21 Ezra Pound, *Guide to Kulchur* (New York: New Directions, 1938), 57-58.

22 Richard P. Feynman, "There's Plenty of Room at the Bottom: An Invitation to Enter a New Field of Physics," *Engineering and Science February* 1960, <http://www.zyvex.com/nanotech.feynman.html>.

23 Susan Stewart, *On Longing: Narratives of the Miniature, the Gigantic, the Souvenir, the Collection* (Baltimore: Johns Hopkins Univ Press, 1984), 43.

24 Drexler, *Engines of Creation,* 221.

25 Ibid., 233.

26 Gregory Benford, "Bio/Nano/Tech," in *Nanodreams,* ed. Elton Elliott (New York: Baen Books, 1995), 193.

27 Kathleen Ann Goonan, "Chicon Live Chat," May 27, 2002, <http://www.cybling.com/chicon/guests/Goonan_Kathleen.html>.

28 David Marusek, "We Were out of Our Minds with Joy," in *Nanotech,* ed. Jack Dann and Gardner Dozois (New York: Ace Books, 1998), 241-42.

29 Drexler, *Engines of Creation,* 172-73.

30 Ian McDonald, *Evolution's Shore* (New York: Bantam Spectra, 1995), 576.

31 Ibid., 598.

32 Alexi Panshin and Cory Panshin, *The World Beyond the Hill: Science Fiction and the Quest for Transcendence* (Los Angeles: Jeremy P. Tarcher, 1989).

33 Kathleen Ann Goonan, "Extending Our Senses," *Locus* 46, no. 6 (2001): 8, 83-84.

34 Alastair Reynolds, "Great Wall of Mars," in *The Year's Best Science Fiction: Eighteenth Annual Collection,* ed. Gardner Dozois (New York: St. Martin's Griffin, 2001), 331.

35 World Transhumanist Association, <http://www.transhumanism.org/> (accessed *2003).*

Chapter 6

1 Don DeLillo, *Cosmopolis* (New York: Scribner, 2003), 85.

2 Larry Smarr, "Nano Space: Microcosmos," *Wired* (June, 2003): 134.

3 Charles Windsor, "Acceptance speech for the Grande Médaille. A speech by the Prince of Wales," Société de Géographie, la Sorbonne, Paris, Thursday 6 February 2003.

4 In "Brave new world or miniature menace? Why Charles fears grey goo nightmare: Royal Society asked to look at risks of nanotechnology," Tim Radford of the UK's *Guardian* newspaper claims that the prince "has raised the spectre of the 'grey goo' catastrophe in which sub-microscopic machines designed to share intelligence and replicate themselves take over and devour the planet." April 29, 2003.

5 Bruno Latour, *We Have Never Been Modern,* trans. Catherine Porter (Cambridge: Harvard University Press, 1993), 6.

6 Ibid., 144.

7 Michael Crichton, *Prey* (New York: HarperCollins, 2002), xiii.

8 Ulrich Beck, *Risk Society: Towards a New Modernity,* trans. Mark Ritter (London: Sage Publications, 1992), 203.

9 Ibid., 58.

10 Anthony Giddens, *The Consequences of Modernity* (Stanford: Stanford University Press, 1990), 39.

11 Ibid., 2.

12 See N. Katherine Hayles, *The Cosmic Web: Scientific Field Models and Literary Strategies in the Twentieth Century* (Ithaca: Cornell University Press, 1984).

13 Rem Koolhaas, "Junkspace," *October* 100 (Spring, 2002): 179.

14 Ralph C. Merkle, "Foresight Debate with *Scientific American*," Foresight Institute, <http://www.foresight.org/SciAmDebate/SciAm/Response.html>, 3.

15 Greg Bear, *Blood Music* (New York: Ace, 1985), 219.

16 Marshall McLuhan, *Understanding Media: The Extensions of Man* (New York: McGraw-Hill, 1964), x.

17 Bruno Latour, "Research Space: The World Wide Lab," *Wired* (June, 2003): 147.

18 Entertainment is one of the primary subsystems which work to produce the "reality" of Niklas Luhmann's *The Reality of the Mass Media:* "This discussion has made plain the special contribution of the 'entertainment' segment to the overall generation of reality. Entertainment enables one to locate oneself in the world as it is portrayed. A second question then arises as to whether this manoeuvre turns out in such a way that one can be content with oneself and with the world" (Stanford: Stanford University Press, 2000), 62.

19 Manuel Castells, *End of Millennium.* Part III of *The Information Age: Economy, Society and Culture* (Oxford: Blackwell Publishers, 1998), 336.

20 Philip S. Anton, Richard Silberglitt, and James Schneider, *The Global Technology Revolution: Bio/Nano/Materials Trends and Their Synergies with Information Technology by 2015,* RAND National Defense Research Institute, public release, 2001, xi.

21 In "The gods of very, very small things," Margaret Wertheim discusses the awkwardness of the NBIC acronym, and repeats a suggestion for a simpler version, BANG: "Bits, Atoms, Neurons and Genes" *The Age,* March 16, 2003.

22 Michel Houellebecq, *The Elementary Particles,* trans. Frank Wynne (New York: Vintage Books, 2000), 15.

23 Crichton, 126.

24 Bill McKibben, *Enough: Staying Human in an Engineered Age* (New York: Times Books, 2003), 92.

25 Beck, 21.

26 Ibid., 11.

27 Neal Stephenson, *The Diamond Age, or, a Young Lady's Illustrated Primer* (New York: Bantam, 1995), 57.

28 Ibid., 43.

29 Crichton, 364.

30 Anton, xvii.

31 National Science Foundation, "Societal Implications of Nanoscience and Nanotechnology," Final Report from the Workshop held at the National Science Foundation (Sept. 28-29, 2000), March 2001.

32 Beck, 56.

33 Anthony Giddens, *Runaway World: How Globalization is Reshaping our Lives* (London: Profile Books, 2002), 35.

34 DeLillo, 79.

35 Beck, 34.

36 Giddens, Consequences, 50-51.

37 Houellebecq, 263.

38 Niklas Luhmann, *Observations on Modernity, trans. William Whobrey* (Stanford: Stanford University Press, 1998), 71.

39 Crichton, 363.

40 Rem Koolhaas, *Wired* (June, 2003), 117.

41 Latour, "Research Space," 147.

42 McLuhan, 3.

43 David Nye, *Narratives and Spaces: Technology and the Construction of American Culture* (Exeter: University of Exeter Press, 1997), 1.

44 Stephenson, 59.

45 Edward W. Soja, "Borders Unbound: Globalization, Regionalism, and the Postmetropolitan Transition" (forthcoming).

46 In William Gibson's *Pattern Recognition* (New York: G.P. Putnam's Sons, 2003), the miniature submarine is a spy device. When Cayce Pollard feels that her border defenses are threatened, she compares that sensation to the way she imagined her father's perimeter-containment job as a Russian spy: "She'd always secretly wanted the KGB spy devices to make it through, because she'd only ever been able to envision them as tiny clockwork brass submarines, as intricate in their way as Faberge eggs. She'd imagined them evading each of Win's snares, one by one, and surfacing in the bowls of staff toilets, tiny gears buzzing" (46). Later, when she suspects that for the first time it is her being followed, the permeable border is her body: "Somewhere, deep within her, surfaces a tiny clockwork submarine" (119).

47 Nicholas Negroponte, *Being Digital* (New York: Alfred A. Knopf, 1996), 6.

48 Ibid., 146.

49 Bear, 71.

50 Houellebecq, 250.

51 Fredric Jameson, *Postmodernism: or, the Cultural Logic of Late Capitalism* (Durham: Duke University Press, 1991), 160.

52 Houellebecq, 223.

53 Koolhaas, "Junkspace," 178. Koolhaas's vision of architecture as junkspace is described by Frederic Jameson as "the new language of space which is speaking through these self-replicating, self-perpetuating sentences" ("Future City," 74). The language of "Junkspace," then, mimics the processes that create Crichton's nanobot swarm in *Prey*.

54 Crichton, 338.

55 Beck, 49.

56 Ibid., 200.

57 Bear, 144.

58 Koolhaas, "Junkspace," 190.

59 Houellebecq, 4.

60 Georges Bataille, "The Notion of Expenditure," *Visions of Excess: Selected Writings, 1927-1939,* trans. Allan Stoekl, with Carl R. Lovitt and Donald M. Leslie, Jr. (Minneapolis: University of Minnesota Press, 1985), 119.

61 Ibid., 123.

62 Houellebecq, 263.

63 In *The Perfect Crime,* Jean Baudrillard asks whether technology is "the lethal alternative to the illusion of the world, or is it merely a gigantic avatar of the same basic illusion, its subtle final twist, the last hypostasis?" (London: Verso, 1996), 5.

64 Jean Baudrillard, *Impossible Exchange,* trans. Chris Turner. (London: Verso, 2001), 33.

65 In the United Kingdom, *The Elementary Particles* is published as *Atomised.* The French title is *Les particules élémentaires.* Both translations are by Frank Wynne.

66 Perhaps this destruction, according to Slavoj Zizek, is exactly what it, and we, need. He challenges anyone who would balk at the consequences of nanotechnology's relative, biogenetic engineering, to welcome their own atomization: "Reducing my being to the genome forces me to traverse

the phantasmal stuff of which my ego is made, and only in this way can my subjectivity properly emerge" ("Bring Me My Phillips Mental Jacket").

Chapter 7

1   Greg Egan, "Dust," in *The Year's Best Science Fiction: Tenth Annual Collection,* ed. Gardner Dozois (New York: St. Martin's, 1993), 87-112; Egan, *Permutation City* (London: Orion, 1994; New York: Harper Prism, 1995).
2   Egan, *Permutation City,* 79
3   Egan, *Permutation City,* 134.
4   Susan E. Lewak, "What's the Buzz? Tell Me What's a-Happening: Wonder, Nanotechnology, and Alice's Adventures in Wonderland," *NanoCulture: Implications of the New Technoscience,* ed. N. Katherine Hayles (Bristol: Intellect Books, 2004), 201-210.
5   Charles N. Brown and William G. Contento, "The Locus Index to Science Fiction (1984-1998)," 2003, <http://www.locusmag.com/index/0start.htm.#TOC>.
6   C.P. Snow, *The Two Cultures and the Scientific Revolution* (New York: Cambridge University Press), 1959.
7   Egan, "Reasons to Be Cheerful," in *The Year's Best Science Fiction: Fifteenth Annual Collection,* ed. Gardner Dozois (New York: St. Martin's, 1998).
8   Brooks Landon, "Less Is More: Much Less Is Much More: The Insistent Allure of Nanotechnology Narratives in Science Fiction Literature," *Nanoculture: Implications of the New Technoscience,* ed. N. Katherine Hayles (Bristol: Intellect Books, 2004), 131-146, Egan, "Axiomatic," in *Nanotech,* ed. Jack Dann and Gardner Dozois (New York: Ace, 1998).
9   Ryman, "A Fall of Angels: Or, On the Possibility of Life in Extreme Conditions," *Unconquered Countries* (New York: St. Martin's, 1994). Although this story was first published in 1994, it was written, according to Ryman, in 1976; "Afterword," *Unconquered Countries,* 273.
10 Ryman, *The Child Garden; or, A Low Comedy* (New York: Tor, 1989).
11 Ryman, "A Fall of Angels," 2.
12 Ursula K. LeGuin, "Introduction," *The Left Hand of Darkness* (New York: Harper & Row, 1976), ix.
13 Ryman, "The Unconquered Country," *Unconquered Countries* (New York: St. Martin's, 1994), 197.
14 K. Eric Drexler, *Engines of Creation: The Coming Era of Nanotechnology* (Garden City, NY: Anchor/Doubleday, 1986).
15 Ryman, "The Unconquered Country," 197.
16 Ryman, *The Child Garden,* 2.
17 Ryman, *Lust,* 56.
18 N. Katherine Hayles, letter to author, 26 August 2003.
19 Kathleen Ann Goonan, *Queen City Jazz* (New York: Tor, 1994), 254.
20 Ryman, *Lust,* 377.

Chapter 8

I would like to thank Kristy Fillmore, Kate Marshall, Brian Rajski, Kenneth Reinhard, Stephen Voyce, and especially Katherine Hayles their advice and support during the writing of this paper and for their comments on the draft.

1 Richard Feynman, "There's Plenty of Room at the Bottom" (1960), in *Nanotechnology Research and Perspectives: Papers from the First Foresight Conference on Nanotechnology,* ed. B.C. Crandall and James Lewis (Cambridge, Mass.: MIT Press, 1992), 360.

2 D.M. Eigler and E.K. Schweizer, "Positioning Single Atoms With a Scanning Tunneling Microscope," *Nature* 344 (1990): 526, 524.

3 Feynman, "There's Plenty of Room at the Bottom," 363.

4 See Ed Regis, *Nano* (Boston: Little, Brown and Co., 1995), 143-147. The sides of the square within which Newman's text was printed were 1/160 mm.

5 Julian C. Chen reviews the limitations of imaging single atoms by field-ion and transmission electron microscopy in *Introduction to Scanning Tunneling Microscopy* (New York: Oxford University Press, 1993), 38-43. For a brief and elementary summary of atomic imaging techniques, see Mark Ratner and Daniel Ratner, *Nanotechnology: A Gentle Introduction to the Next Big Idea* (Upper Saddle River, N.J.: Prentice Hall, 2003), 39-43.

6 Eigler, "From the Bottom Up: Building Things With Atoms," *Nanotechnology,* ed. Gregory Timp (New York: Springer-Verlag, 1999), 425-435.

7 For a technical introduction to the STM, see Chen. Zhong Lin Wang, Yi Lui, and Ze Zhang offer an up-to-date overview of the STM's operation and applications in *Handbook of Nanophase and Nanostructured Materials,* vol. 2 (New York: Kluwer Academic/Plenum Publishers, 2003). Perhaps the best description for the general reader is provided by Charles Lieber, "Scanning Tunneling Microscopy," *Chemical & Engineering News* (18 Apr. 1994): 28-43, but see also Eigler, "From the Bottom Up" and Jim Gimzewski and Victoria Vesna, "The Nanomeme Syndrome: Blurring of fact and fiction in the construction of a new science," *Technoetic Arts: a journal of speculative research* 1, no. 1 (2003). The Institute für Allgemeine Physik offers a slideshow at <http://www.iap.tuwien.ac.at/www/surface/STM_Gallery/stm_schematic.html>, from which the schematic above was gleaned. Pivotal articles on atomic positioning include J.S. Foster, J.E. Frommer, and P.C. Arnett, "Molecular Manipulation Using a Scanning Tunneling Microscope," *Nature* 331 (1988): 324-326 and Eigler and Schweizer, "Positioning Single Atoms with a Scanning Tunneling Microscope."

8 On the atomic switch, see D.M. Eigler, C.P. Lutz, and W.E. Rudge, "An Atomic Switch Realized with the Scanning Tunneling Microscope," *Nature* 352 (1991): 600-602. On the molecular abacus, see M.T. Cuberes, R.R. Schittler, and J.K. Gimzewski, "Room-Temperature Repositioning of C60 Molecules at Cu Steps: Operation of a Molecular Counting Device," *Applied Physics Letters* 69, no. 20 (1996): 3016-3018. On quantum corrals, see M.F. Crommie, C.P. Lutz, and D.M. Eigler, "Imaging Standing Waves in a Two Dimensional Electron Gas," *Nature* 363 (1993): 524-527; and Eigler, "From the Bottom Up." For a concise introduction to "quantum constructions" see Mark Reed, "Quantum Constructions," *Science* 262 (1993): 195.

9 N. Katherine Hayles, *Writing Machines* (Cambridge, Mass.: MIT Press, 2002), 24.

10 Lisa Gitelman, *Scripts, Grooves, and Writing Machines: Representing Technology in the Edson Era* (Stanford: Stanford University Press, 1999), 11.

11 Emily Dickinson, n. 640, *The Complete Poems of Emily Dickinson,* ed. Thomas H. Johnson (Boston: Little, Brown and Co., 1960).

12 Roman Jakobsen, *Selected Writings,* vol. II (The Hague: Mouton, 1971), 132.

13 Emile Benveniste, *Problems in General Linguistics,* trans. Mary Elizabeth Meek (Coral Gables, Fla.: University of Miami Press, 1971), 219.

14 G.F.W. Hegel, *Phenomenology of Spirit,* trans. A. V. Miller (Oxford: Oxford University Press, 1977), 60.

15 Roland Barthes, *The Rustle of Language,* trans. Richard Howard (New York: Hill and Wang, 1986), 51. For a summary of theories of deixis–to which "I" am indebted "here"–see Jed Rasula, "The Poetics of Embodiment: A Theory of Exceptions" (Ph.D. diss., University of California at Santa Cruz, 1989), 62-78.

16 Benveniste, *Problems in General Linguistics,* 224, 225.

17 Steve McCaffery, *Shifters* (1976), in *Seven Pages Missing: Volume One* (Toronto: Coach House Press, 2000), 127.

18 McCaffery, "Shifters: a note" (1976), in *Seven Pages Missing,* 450.

19 Qtd. in Julia Kristeva, *Revolution in Poetic Language,* trans. Margaret Waller (New York: Columbia University Press, 1984), 256.

20 Dickinson, *The Manuscript Books of Emily Dickinson,* ed. R.W. Franklin, vol. 2 (Cambridge, Mass.: Belknap Press, 1981), 797.

21 On the function and semiotic value of the dash in Dickinson's poetry, see, in particular, Geoffrey Hartman, Criticism in the *Wilderness: The Study of Literature Today* (New Haven: Yale University Press, 1980), 126; Susan Howe, *My Emily Dickinson* (Berkeley: North Atlantic, 1985), 23; and Christanne Miller, *Emily Dickinson: A Poet's Grammar* (Cambridge, Mass., Harvard University Press), 53. For detailed readings of Poem #640, see Susan Cameron, *Choosing Not Choosing: Dickinson's Fascicles* (Chicago: Chicago University Press, 1992); Judith Farr, *The Passion of Emily Dickinson* (Cambridge, Mass.: Harvard University Press, 1992); Gary Hawkins, "Constructing and Residing in the Paradox of Dickinson's Prismatic Space," *The Emily Dickinson Journal* IX, no. 1 (2000): 49-70; Suzanne Juhasz, *The Undiscovered Continent: Emily Dickinson and the Space of the Mind* (Bloomington: Indiana University Press, 1983); Inder Nath Kher, *The Landscape of Absence* (New Haven: Yale University Press, 1974); Denise Kohn, "'I cannot live with You –'," *An Emily Dickinson Encyclopedia,* ed. Jane Donahue Eberwein (Westport, Conn.: Greenwood Press, 1998), 154-5; Gary Lee Stonum, *The Dickinson Sublime* (Madison, Wis.: University of Wisconsin Press, 1990); and Cynthia Griffen Wolff, *Emily Dickinson* (Reading, Mass.: Addison Welsley, 1988). So far as I can tell, the features of the line that I have focused on here have not been discussed in the criticism.

22 Emmanuel Levinas, *Totality and Infinity: An Essay on Exteriority,* trans. Alphonso Lingis (Pittsburgh: Duquesne University Press, 1981), 43.

23 As an approach to Dickinson's poem, the Levinasian face to face significantly qualifies the relation of "I" to "you" described by Benveniste. While the latter conceives of the self/other relation as symmetrical (a "polarity of persons" in which the other "becomes my echo to whom I say *you* and who says *you* to me"), the formal structure articulated by Levinas is fundamentally asymmetrical (the You is prior to, higher than, master of, sovereign over the I). In *Otherwise than Being, Or Beyond Essence,* trans. Alphonso Lingis (Pittsburgh: Duquesne University Press, 1981), Levinas goes so far as to describe the self as held-hostage-by the Other–a formulation that is far from incompatible with Dickinson's masochistic sensibility. According to Levinas, it is "through the condition of being hostage that there can be in the world pity, compassion, pardon and proximity–even the little that there is, even the simple 'After you, sir'" (OTB 117). In the modulation of "I cannot live with You" into "You there – I – here –", it is precisely such consideration that enables a

meeting apart: after "You", " – I – here –" is revealed as *tied* to the domain of the Other–hostage of and hosted by "there." I am indebted to Brian Rajski for pressing the distinction between Levinas and Benveniste (personal correspondence).

24 To do so is to read Levinas with Derrida.

25 Robert Smithson, *Robert Smithson: The Collected Writings,* ed. Jack Flam (Berkeley: University of California Press, 1996), 96.

26 Craig Dworkin, *Reading the Illegible* (Evanston, Ill.: Northwestern University Press, 2003), 5. By "radical" formalism, Dworkin means to specify an attention to the politics of form, and hence to the historical materiality of the signifier. He adapts the term from Bruce Andrews's definition of a "radical praxis," which "involves the rigors of formal celebration, a playful infidelity, a certain illegibility within the legible: an infinitizing, a wide-open exuberance, a perpetual motion machine, a transgression" (qtd. in Dworkin, 5). The kind of radical formalism practiced by Dworkin, along with critics/poets like Andrews, Christian Bök, Marjorie Perloff, Michael Davidson, Charles Altieri, Susan Howe, Lyn Hejinian, Steve McCaffery, Bob Perelman, and Jed Rasula (to name a few) amounts to an investigation of the political science of the poem.

27 Eigler, "From the Bottom Up," 425

28 *OED.*

29 Martin Heidegger, "Poetically Man Dwells," in *Poetry, Language, Thought,* trans. Albert Hofstadter (New York: HarperCollins, 1971), 225.

30 Mark Hansen, *Embodying Technesis: Technology Beyond Writing* (Ann Arbor: Michigan University Press, 2000), 4 (hereafter cited in the text as *ET).*

31 Hansen asks, for example, if Derrida's "restriction of exteriority (including the exteriority of matter itself) [does] not already decide the problem of technology in advance and according to a dogmatic philosophical program? And if so, does it not thereby compromise *from the start* the radicality of Derrida's grammatological circumscription of cybernetics–both by instituting in general a *nouo-centric* origin and by preserving all of those metaphysical concepts (soul, life, value, choice, and especially memory) that, as Derrida puts it, serve 'to separate the machine from man' (9)" (127). If this doesn't exactly sound like Derrida, we have only to remember page nine of the *Grammatology* to see why: "If the theory of cybernetics is by itself to oust all metaphysical concepts–including the concepts of soul, of life, of value, of choice, of memory–*which until recently have served to separate the machine from man,* it must conserve the notion of writing, trace, grammé [written mark] or grapheme, until its own historico-metaphysical character is also exposed" (Jacques Derrida, *Of Grammatology,* trans. Gayatri Spivak [Baltimore: Johns Hopkins University Press, 1974], 9, emphasis added). As Derrida puts it in *Positions:* "what I denounce is attributed to me, as if one were in less of a hurry to criticize or discuss me, than first to put oneself in my place in order to do so" (53). Only through such contortions could Derrida be represented as a *"nouo-centric"* humanist. For what seems to me a more measured assessment of Derrida's relevance to science studies, see Timothy Lenoir's introduction to *Inscribing Science* (Stanford: Stanford University Press, 1998). As for Heidegger's reduction of technology to the status of instrument, he is also accused of that which he denounces in terms that could not be more clear: "Meanwhile, man, precisely as the one so threatened, exalts himself and postures as lord of the earth. In this way the illusion comes to prevail that everything man encounters exists only insofar as it is his construct" ("The Question Concerning Technology," in *Basic Writings,* trans. William Lovitt [New

York: HarperCollins, 1993], 332). And five pages later: "So long as we represent technology as an instrument, we remain transfixed in the will to master it" (337). Heidegger makes it equally clear that his consideration of enframing as an mode of "revealing" does *not* constitute a subordination of modern technology to *Dasein:* "Where do we find ourselves if now we think one step further regarding what enframing itself actually is? It is nothing technological, nothing on the order of the machine. It is the way in which the actual reveals itself as standing-reserve. Again we ask: Does such revealing happen somewhere beyond all human doing? No. But neither does it happen exclusively in man, or definitively *through* man" (ibid., 328-9).

32 Hansen, *Embodying Technesis,* 262.

33 Ibid., 148.

34 Heidegger, "The Question Concerning Technology," 337.

35 Heidegger, "...Poetically Man Dwells...," 213.

36 K. Eric Drexler, "Machines of Inner Space" (1989), in *Nanotechnology Research and Perspectives: Papers from the First Foresight Conference on Nanotechnology,* ed. B.C. Crandall and James Lewis (Cambridge, Mass.: MIT Press, 1992), 326.

37 Steve McCaffery, *Prior to Meaning: The Protosemantic and Poetics* (Evanston, Ill.: Northwestern University Press, 2001), xv.

38 On the relevance of quantum theory to modernist poetry and poetics, see Daniel Albright, *Quantum Poetics: Yeats, Pound, Eliot, and the Science of Modernism* (Cambridge: Cambridge University Press, 1997).

39 McCaffery, *Prior to Meaning,* xv.

40 Ibid, xxii.

41 Ibid, xix.

42 *Carnival: The Second Panel, 1970-1975* can be viewed online at the Coach House Books website: <http://www.chbooks.com/online/carnival/index.html>.

43 McCaffery, *Seven Pages Missing,* 446.

44 Dworkin, *Reading the Illegible,* 72.

45 Rasula, "The Poetics of Embodiment," 476. Rasula's exemplary performance of wreading is *This Compost: Ecological Imperatives in American Poetry.*

46 McCaffery, "Knowledge Never Knew" (1983), *Seven Pages Missing,* 204.

47 McCaffery explains in his introduction to *Carnival* that, "having discovered, explored, and tested the parameters of the typewriter in *The First Panel, The Second Panel* places the typed mode in agonistic relation with other forms of scription: xerography, xerography with xerography (i.e. metaxerography and disintegrative seriality), electrostasis, rubber stamp, tissue texts, hand-lettering and stencil" (*Seven Pages Missing,* 445).

48 On quasi-objects, situated in the "Middle Kingdom" between nature and society, see Bruno Latour, *We Have Never Been Modern,* trans. Catherine Porter (Cambridge, Mass.: Harvard University Press), 51. Images of "atoms" produced by the STM are, of course, nothing more than the digitally processed pattern of the tip's movement over the sample, and bear nothing but a scalar resemblance to the "matter" under investigation.

49 Maurice Merlau-Ponty, *The Visible and Invisible,* trans. Alphonso Lingis, ed. Claude Lefort (Evanston, Ill.: Northwestern University Press, 1968), 144.

50 Neil Postman, *Technopoly: The Surrender of Culture to Technology* (New York: Vintage, 1992).

51 Leiber, "Scanning Tunneling Microscopy," 28.

52 Qtd. in Lew Phelps, "Nanotechnologist wins Nobel Prize in chemistry for discovering Fullerenes" (March 1997), *Foresight Institute*, <http://www.islandone.org/Foresight/Updates/Update27/Update27.1.html>.

53 Christian Bök, *Crystallography* (1994) 2nd ed. (Toronto: Coach House, 2003), 156 (hereafter cited in the text as C).

54 Bök, *'Pataphysics: The Poetics of an Imaginary Science* (Evanston, Ill.: Northwestern University Press, 2002), 4.

55 Caroline Bergvall, *Goan Atom* (San Francisco: Krupskaya, 2001), 7 (hereafter cited in the text as *GA*).

56 Drew Milne, "A Veritable Dollmine," *Jacket Magazine* (July 2000), <http://jacketmagazine.com/12/milne-bergvall.html>.

57 Lieber, "Scanning Tunneling Microscopy," 39.

58 French inf(l)ects *Goan Atom* as a dangerous supplement in the most capaciously Derridean sense, since it is Bergvall's "original" language.

59 Jean Baudrillard, *The Ecstasy of Communication*, trans. Bernard and Caroline Schutze (New York: Semiotext(e), 1988), 43.

60 Commissioned as a sound text installation for Root Festival, at Hull Time Based Arts in November, 1999, *Ambient Fish* is available as a net-text at <http://epc.buffalo.edu/authors/bergvall/amfish/cbflash.html>.

61 Drexler, "Machines of Inner Space," 344-45.

62 Ibid, 325.

63 Thomas Pynchon, *Gravity's Rainbow* (1973) (New York: Penguin, 2000), 726.

64 We do indeed arrive at the limits of fabrication when we find the logo described on the IBM Research website as an artist's "way to give something back" to his patron: <http://www.almaden.ibm.com/vis/stm/atomo.html#stm10>.

65 George Oppen, "Route" (1968), in *George Oppen: New Collected Poems*, ed. Michael Davidson (New York: New Directions, 2002), 202.

Chapter 9

1 Nicholas Negroponte. *Being Digital* (New York: Alfred A. Knopf, 1996), 14. Manovich identifies modularity, automation, variability, and transcoding as the primary characteristics of new media, modularity and transcoding being the two that enable variability and automation in *Language of New Media* (Cambridge: MIT Press, 2001).

2 The nano scale is in the range of 1-100 nanometers; a nanometer is one billionth of a meter. Nanotechnology, often used interchangeably with nanoscience, is the construction and control of materials or systems at this scale. Biotechnology and genetics, fields intimately related to nanoscience and nanotechnology, are entirely dependent on the transcoding of organic into binary code. The output of such digital imaging of nanoscience experiments comes in the form of linear graphs or colorful molecular illustrations.

3 See Kate Marshall's essay in this collection, 79.

4 See for example Gayle L. Ormiston and Raphael Sassower, editors, *Narrative Experiments: The Discursive Authority of Science and Technology* (Minneapolis: University of Minnesota Press, 1990) and Joseph Margolis, *Texts Without Referents: Recovering Science and Narrative* (Persistence of Reality, Vol. 3) (London: Blackwell, 1989).

5 Quoted in Regis, 23.

6 Quoted in Ed Regis, *Nano: The Emerging Science of Nanotechnology: Remaking the World-Molecule by Molecule* (New York: Little, Brown, and Company, 1995), 23.

7 Quoted in Sue Vorenberg, "Nanoscience: Big Interest in Studying the Very Small," *National Geographic News Online* (August 22, 2002) <http://news.nationalgeographic.com/news/2002/08/0822_020822_nanoscience.html> (accessed October 15, 2003).

8 Eric K. Drexler. *Engines of Creation: The Coming Era of Nanotechnology.* (New York: Anchor Books, 1986), 296.

9 Quoted in Regis, 194.

10 For a discussion of the STM as a writing machine or instrument of inscription, see Nathan Brown's "Needle on the Real: Technoscience and Poetry at the Limits of Fabrication" *infra* 173-190.

11 Margaret Wertheim. "Buckyballs and Screaming Cells: The Amazing Miniature World of UCLA Chemist Jim Gimzewski," *LA Weekly* (April 4-10, 2003), 29, <http://www.laweekly.com/ink/03/20/features-wertheim.php> (accessed October 15, 2003). Colin Milburn in "Nanotechnology in the Age of Posthuman Engineering: Science Fiction as Fiction" *infra* 109-130, explores at length the connection between nanoscience and science fiction literature.

12 Samuel Taylor Coleridge, *Biographia Literaria* (1817), Ch. XIV.

13 Stian Grogaard, "Low Tech-High Concept: Digital Media, Art, and the State of the Arts" in *Digital Media Revisited,* edited by Gunnar Liestol, Andrew Morrison, and Terje Rasmussen (Cambridge: MIT Press, 2003), 273.

14 <http://www.wtec.org/loyola/nano/NSET.Societal.Implications> (accessed October 15, 2003).

15 <http://nint-innt.nrc-cnrc.gc.ca/home.html> (accessed October 15, 2003). Gregory Timp, "Nanotechnology" in *Nanotechnology,* edited by Gregory Timp (New York: Springer Verlag 1999), 1. Timp's rhetorical language continues, presenting nanotechnology as a "world [that] is even more exotic than Marco Polo's Asia seemed to be, since the classical laws of physics that govern the mechanics of our common experience are suspended on this frontier."

16 Richard P. Feynman. "There's Plenty of Room at the Bottom: An Invitation to Enter a New Field of Physics," *Engineering and Science* (February, 1950).

17 Friedrich Kittler. *Gramophone, Film, Typewriter,* translated by Geoffrey Winthrop-Young and Michael Wurtz (Palo Alto: Stanford University Press, 1999, 1986), xxxix. In the chapter titled "Typewriter," Kittler applies McLuhan's argument that the typewriter "fuses composition and publication, causing an entirely new attitude to the written and printed word" (McLuhan 260).

18 Atoms are so small that they cannot be detected by light and therefore are not perceptible to the human eye or an optical microscope; to be visible, an object must be large enough to catch and reflect light in the cross section of a lightwave, which is approximately 500 nanometers wide.

19 A delineation of the STM's reading process explains the conversion of sensory modes and types of information. A conducting needle at the tip of the microscope nearly touches the atom's surface, stimulating a tunneling current to flow between the gap of the microscope's tip and atom's surface. As the needle scans the atom's surface, it registers variations in the intensity of the (analog) current and plots the depth of the topological surface according to the strength of the current. A digital-analog converter then translates these detectable energy levels into binary code (digital) and uses this code to create an (analog) visual representation of the invisible atom, either in the form of a graph or a topological illustration. The tunneling currents are translated into an image

of the atomic surface by designating a vertical height, color, or brightness to each sensory scan and then assigning the position of a light-source. "The shadowing effect on one side of the pinned molecule is an electronic artefact [sic] related to the way in which the data are acquired." J.S. Foster, et al. "Molecular Manipulation Using a Tunneling Microscope," *Nature.* Vol. 331 (January 28, 1988): 324.

20 Friedrich Kittler, "There is No Software," Ctheory.net, <http://www.ctheory.net/text_file.asp?pick=74> (accessed October 15, 2003).

21 John Cayley. "The Code Is Not the Text (unless it is the Text)" *Electronic Book Review* (September 10, 2002): 3, <http://www.electronicbookreview.com/v3/servlet/ebr?command=view_essay&essay_id=cayleyele> (accessed October 15, 2003).

22 I use the "happens" to invoke the "Happenings" produced and defined by Allan Kaprow in the 1950s and 1960s. Happenings inaugurated a different perspective for presenting and understanding performance. For a connection between Happenings and new media, see "'Happenings' in the New York Scene" (Ch. 6) in *The New Media Reader,* edited by Noah Wardrip-Fruin and Nick Montfort (Cambridge: The MIT Press, 2003), 83-88.

23 Cayley, 3.

24 N. Katherine Hayles, "Translating Media: Why We Should Rethink Textuality," *The Yale Journal of Criticism* (Vol.16, No. 2, Fall 2003).

25 Jacques Derrida. "Structure, Sign and Play in the Discourse of the Human Sciences," *Writing and Difference,* translated by Alan Bass (Chicago: University of Chicago Press, 1978), 292.

26 Peter Lunenfeld. "Unfinished Business," *The Digital Dialectic: New Essays on New Media* (Cambridge: MIT Press, 1999), 14.

27 Pierre Levy, *Cyberculture,* translated by Robert Bononno (Minneapolis: University of Minnesota Press, 1997), 45; Lunenfeld, 7.

28 Rita Raley, "Reveal Codes: Hypertext and Performance," *Postmodern Culture.* Vol. 12, No. 1 (2001) paragraph 10, <http://muse.jhu.edu/journals/pmc/v012/12.1raley.html> (accessed October 15, 2003).

29 Brenda Laurel, *Computers as Theater* (Reading: Addison-Wesley Publishing Company, 1993), 7.

30 Peter Lunenfeld, *Snap to Grid: A User's Guide to Digital Arts, Media, and Cultures* (Cambridge: MIT Press, 2000), 37.

31 Richard Schechner discusses the influence of the Uncertainty Principle on the arts: "Arts, once the home of strict choreography, precise scores, and fixed mise-en-scenes, were opened to change processes, unpredictable eruptions from the unconscious, and improvisation" in *Performance Studies, An Introduction* (Routledge, 2002), 99.

32 Raley, paragraph 14.

33 Strehle, 9.

34 Feynman, <http://www.zyvex.com/nanotech/feynman.html> (accessed October 15, 2003).

35 Strehle, 9. Max Planck proved (in 1901) that the energy of electron orbits is quantized, occurring in discrete, not continuous, amounts.

36 W.J.T. Mitchell. "Spatial Form in Literature: Toward A General Theory," *The Language of Images,* edited by W.J.T. Mitchell (Chicago: University of Chicago Press, 1974), 288.

37 See Roger Schank's *Dynamic Memory* (1982).

38 See the quotation from Kittler on page 312 of this essay.

39 Katherine Hayles, *Writing Machines.* (Cambridge: MIT Press, 2002), 117.

40 See page 316 of this essay.

41 To highlight one example, the single sentence of text that comprises chapter one is portrayed into two very distinct and divergent interpretations. In View Text the question "Is this what you wanted?" is repeated throughout the vertical register. The words flicker against a red background with staccato sounds whose acceleration heightens the anxiety of obsessive questioning. By contrast, in Perform Text the question is spoken by a soft, sexy voice that blends into the music while a shifting arrangement of square photographs, snapshots depicting a heterogeneous array of people, slide across a back and white background. The tone here is calm, even seductive. This is a drastically different portrayal than View Text's feeling of being trapped within a mind, a screen, and an endlessly repeated sentence. Chroma thus demands that the reader engage both sections—to View and Perform the text. The result is a presentation of digital textuality as literature and performance, a merger that transforms our understanding of what literature is and how we approach it.

42 Lev Manovich titles our digital culture "the culture of the remix," for, "in computer culture, authentic creation has been replaced by selection from a menu," 124.

43 Katherine Hayles discusses the "flickering signifier" in chapter two of *How We Became Posthuman: Virtual Bodies in Cybernetics, Literature, and Informatics* (Chicago: University of Chicago, 1999), 25-49.

44 Michael Joyce, *Of Two Minds: Hypertext Pedagogy and Poetics* (Ann Arbor: University of Michigan Press, 1995), 233.

45 Friedrich Kittler. *Discourse Networks, 1800/1900,* translated by Michael Metteer with Chris Cullens (Palo Alto: Stanford University Press, 1990), 369.

46 The series includes *The Lair of the Marrow Monkey* (1998), which is the prequel to *Chroma,* and "The Institute for Investigations into the Mind of Marrow" (1997) which only exists online in an excerpt (as of February 2004).

47 Jane Yellowlees Douglas examines the concept of literary closure and its application to electronic literature in *The End of Books—Or Books without End?: Reading Interactive Narratives* (Ann Arbor: University of Michigan Press, 2000); Joyce, 85.

48 <http://www.marrowmonkey.com/chroma/07.html> (accessed October 15, 2003).

49 Lunenfeld "Unfinished," 7; Lunenfeld "Screen," 14.

50 Strehle, 13.

Chapter 10

1 Andrew Lloyd Webber and Tim Rice, *What's the Buzz/Strange Thing Mystifying,* "*Jesus Christ Superstar,*" Decca, DXSA 7206, DL 79178, DL 79179 (3 discs), sound recording, 1970.

2 "Episode 19: Hit a Sista Back," James Cameron's *Dark Angel: The Complete First Season,* DVD, directed by James Whitmore, Jr. (2001; Beverly Hills, CA: Twentieth Century Fox, 2003).

3 Nanotechnology relates to the application of devices that use structures having at least one dimension of about one to several hundred nanometers.

4 Wonder: "the object of astonishment . . . an event brought about by miraculous or supernatural power; a miracle." *Oxford English Dictionary* online, s.v. "Wonder," <http://dictionary.oed.com/cgi/entry/00286810?query_type=word&queryword=wonder&edition=2e&first=1&max_to_show=10&sort_type=alpha&result_place=1&search_id=b2t6-Q7dMSC-202&hilite=00286810> (accessed June 20, 2003).

5   Michael L. Roukes, "Foreword," *Understanding Nanotechnology: From the Editors of Scientific American,* comp. Sandy Fritz (New York: Warner Books, 2002) viii.

6   Lewis Carroll was the pen name of Oxford don and noted mathematician Charles Dodgson. Dodgson was as dull and ordinary as Carroll was imaginative and outrageous. He would spend much of his life denying that there was a connection between the two men.

7   *Alice's Adventures in Wonderland* is a popular metaphor for nanotechnology as demonstrated through the following passages:

"The realm of nanotechnology exists on a scale in between the subatomic realm of elemental particles and the macroscopic world of electronic devices and circuits. 'To understand nanotechnology,' says [Professor Alexander] Balandin, 'you have to realize that as you reduce the scale, at some point quantitative changes become qualitative changes.' This is an Alice-in-wonderland world that plays by its own rules—rules researchers have yet to understand fully." "Exploring the World at Nanoscale," *The National Science Foundation at the University of California* (summer 2001), <http://www.ee.ucr.edu/~alexb/uc-nano.pdf>.

"Under the pen name of Lewis Carroll, Oxford mathematician Charles Dogsdon wrote about a world that lay . . . down the rabbit hole in his classic tale, *Alice in Wonderland.* At times, Alice became so small that she collapsed like a telescope and so large that she had to stick a leg up a chimney to fit in a house. When scientists go down the 'rabbit hole' of nanotechnology, they encounter a similar Alice-in-Wonderland surrealistic world." Steve Tally, "A Shrinking World Inside Agriculture," *Purdue Agricultures Magazine,* (summer 2002), <http://www.agriculture.purdue.edu/agricultures/past/summer2002/features/feature_01_p2.html>.

"Turn your minds if you will to a world in miniature. We're talking about nanotechnology, a discipline which involves manipulating material on a level so minute it's difficult to comprehend. But that's exactly what technologists the world over are trying to do. Like *Alice in Wonderland,* they're searching for that magic elixir, to open doors into a world where anything is possible," Richard Taylor, "Nanotech Time," *British Broadcasting Corporation,* (August 9, 2003), <http://www.bbcworld.com/content/clickonline_archive_32_2001.asp?pageid=666&co_pageid=2>.

"This may be remembered as the 'Alice in Wonderland' decade for new technology. More and more businesses are moving into the world of nanotechnology, where particles of common materials are shrunk to such a minuscule size that they behave in unexpected–and often useful–ways." Barnaby J. Feder, "In the World of the Very Small, Companies Make Big Plans," *New York Times,* (December 16, 2002), sec. C19.

"This tiny realm is an Alice-in-Wonderland world, where materials behave in unexpected ways and can be put together to do amazing new things," Rose Simone, "UW Team Takes Big Steps in a Truly Tiny World: Nanotechnology Research has the Potential to Shape Our Future," *The Record,* (October 24, 2002), <http://therecord.com/business/techspot/business_techspot_02102315533.html>.

8   Gary Stix notes that "With recognition has come lots of money—lots, that is for something that isn't a missile shield. The National Nanotechnology Initiative (NNI) initiated by President Bill Clinton is a multi-agency program intended to provide a big funding boost to nanoscience and engineering. The $422-million budget in the federal fiscal year 2000 marked a 56 percent jump in nano spending from a year earlier. . . Nano mania flourishes everywhere. In 2001, more than 30 nanotechnology research centers and interdisciplinary groups sprouted in universities; fewer

than 10 existed in 1999," "Little Big Science," *Understanding Nanotechnology: From the Editors of Scientific American,* ed. Sandy Fritz (New York: Warner Books, 2002), 6-16, especially 7.

For an in-depth discussion of the role of former Vice President Al Gore's fundamental role in involving the federal government in nanotech research, see William Illsey Atkinson, Nanocosm: Nanotechnology and the Big Changes Coming from the Inconceivably Small (New York: AMACOM/ American Management Association, 2003), 86-88.

For more statistics on business and governmental spending in nanotechnology, see William Illsey Atkinson, *Nanocosm: Nanotechnology and the Big Changes Coming from the Inconceivably Small* (New York: AMACOM/American Management Association, 2003), 58-9 and Mark Ratner and Daniel Ratner, *Nanotechnology: A Gentle Introduction to the Next Big Idea,* (Upper Saddle River, NJ: Prentice Hall, 2003), 2-3.

9   Michael Roukes, "Plenty of Room, Indeed," *Understanding Nanotechnology: From the Editors of Scientific American,* ed. Sandy Fritz (New York: Warner Books, 2002), 26.

10 Jessica Pressman. "Nano Narrative: A Parable from Electronic Literature." *NanoCulture: Implications of the New Technoscience,* ed. N. Katherine Hayles (Bristol: Intellect Books, 2004), 202.

11 Brooks Landon. "Less is More: Much Less is Much More: The Insistent Allure of Nanotechnology Narratives in Science Fiction," *NanoCulture: Implications of the New Technoscience,* ed. N. Katherine Hayles (Bristol: Intellect Books, 2004), 134.

12 Colin Milburn,"Nanotechnology in the Age of Posthuman Engineering: Science Fiction as Science," NanoCulture: Implications of the New Technoscience, ed. N. Katherine Hayles (Bristol: Intellect Books, 2004), 120.

13 Steven Swann Jones, *The Fairy Tale: The Magic Mirror of Imagination* (New York: Routledge, 2002), 8-9.

14 Kate Marshall. "Atomizing the Risk Technology," *NanoCulture: Implications of the New Technoscience,* ed. N. Katherine Hayles (Bristol: Intellect Books, 2003), *infra* 147-160.

15 Jack David Zipes, *The Trials & Tribulations of Little Red Riding Hood,* 2nd ed. (New York: Routledge, 1993), 53.

16 For a detailed discussion of the development of children's literature from the oral tradition in 19th-century England, see Jack Zipes, "Introduction." *Victorian Fairy Tales: The Revolt of the Fairies and Elves* (New York: Methuen, 1987), xiii-xxix.

17 Peter Hunt, *Children's Literature: An Anthology, 1801-1902* (Malden: Blackwell, 2001), 240.

18 Colin Manlove reinforces the idea that Wonderland is supported by rules rather than chaos when he states, "the weirdness of Wonderland consists not in the wholesale invention of unfamiliar things but in the re-ordering of a familiar universe, so that things exist in strange relationship to one another," "Victorian and Modern Fantasy: Some Contrasts," *The Celebration of the Fantastic: Selected Papers from the Tenth Anniversary International Conference on the Fantastic in the Arts,* edited by Csilla Bertha Marshall B. Tymn and Donald E. Morse, (Westport: Greenwood Press, 1992), 43.

19 As an Oxford don, Charles Dodgson was not allowed to marry and was perceived as too shy and retiring to keep company with women his own age. For a discussion of his interest in "child-friends," young pre-adolescent girls whom he often photographed, see Jo Elwyn Jones and J. Francis Gladstone, *The Alice Companion: A Guide to Lewis Carroll's Alice Books* (Houndmills, Basingstoke, Hampshire and London: Macmillan Press LTD, 1998), 38-9. For a discussion of Alice Liddel, see Jones and Gladstone, *The Alice Companion: A Guide to Lewis Carroll's Alice Books,* 157-161. For a discussion of his influence on Nabokov, who "was inspired by Alice to write the

novel, Lolita," see Jones and Gladstone, *The Alice Companion: A Guide to Lewis Carroll's Alice Books,* 185.

20 Brian Attebery, "Dust, Lust, and Other Messages from the Quantum Wonderland," *NanoCulture: Implications of the New Technoscience,* ed. N. Katherine Hayles (Bristol: Intellect Books, 2004), 161.

21 Adriana de Souza e Silva, "The Invisible Imaginary: Museum Spaces, Hybrid Reality and Nano-technology," *NanoCulture: Implications of the New Technoscience,* ed. N. Katherine Hayles (Bristol: Intellect Books, 2004), 28.

22 Karen Barad, "Agential Realism: Feminist Interventions in Understanding Scientific Practices," *The Science Studies Reader,* ed. Mario Biagioli (New York: Routledge, 1999), 3.

23 Karen Barad, "Agential Realism," Encyclopedia of Feminist Theories, ed. Lorraine Code (New York: Routledge, 2000), 15.

24 Lewis Carroll, *Alice in Wonderland,* 2nd ed. (New York: W.W. Norton & Co., 1992), 7. Hereafter cited in text.

25 N. Katherine Hayles, "Introduction," *NanoCulture: Implications of the New Technoscience,* ed. N. Katherine Hayles (Bristol: Intellect Books, 2004), 18.

○○○

Bibliography

Albright, Daniel. *Quantum Poetics: Yeats, Pound, Eliot, and the Science of Modernism*. Cambridge: Cambridge University Press, 1997.

Alcor Life Extension Foundation. <http://www.alcor.org> (accessed 2003).

Alivisatos, A. Paul. "Less Is More in Medicine." In Fritz, comp., *Understanding Nanotechnology*, (2002): 56-71.

Amato, Ivan. "Scanning Probe Microscopes Look into New Territories." *Science* 262 (October 8, 1993).

Anderson, Kevin J., and Doug Beason. *Assemblers of Infinity*. New York: Bantam Spectra, 1993.

Anton, Philip S., Richard Silberglitt, and James Schneider. *The Global Technology Revolution: Bio/ Nano/Materials Trends and Their Synergies with Information Technology by 2015*. RAND National Defense Research Institute, public release, 2001.

Ashley, Steven. "Nanobot Construction Crews." In Fritz, comp., *Understanding Nanotechnology*, 86-91.

Astumian, R. Dean. "Making Molecules into Motors." In Fritz, comp., *Understanding Nanotechnology*, 72-86.

Atkins, P.W. *The Second Law: Energy, Chaos, and Form*. New York: W.H. Freeman & Co, 1994.

Atkinson, William Illsey. *Nanocosm: Nanotechnology and the Big Changes Coming from the Inconceivably Small*. New York: AMACOM/American Management Association, 2003.

Baake, Ken. *Metaphor and Science: The Challenge of Writing Science*. Suny Series: Studies in Scientific and Technical Communication. New York: State University of New York Press, 2003.

Ballard, J.G. *The Crystal World*. New York: Farrar, Straus & Giroux, 1966.

Barad, Karen. "Agential Realism: Feminist Interventions in Understanding Scientific Practices." In *The Science Studies Reader,* edited by Mario Biagioli, 1-11. New York: Routledge, 1999. In *Encyclopedia of Feminist Theories*, edited by Lorraine Code, 15-16. New York: Routledge, 2000.

Barthes, Roland. *The Rustle of Language*. Translated by Richard Howard. New York: Hill and Wang, 1986.

Bataille, Georges. "The Notion of Expenditure." In *Visions of Excess: Selected Writings, 1927-1939*, 116-29. Translated by Allan Stoekl with Carl R. Lovitt and Donald M. Leslie, Jr. Minneapolis: University of Minnesota Press, 1985.

Baudrillard, Jean. *Écran Total*. Paris: Éditions Galilée, 1997.

——. *The Ecstasy of Communication*. Translated by Bernard and Caroline Schutze. New York: Semiotext(e), 1988.

——. *Impossible Exchange*. Translated by Chris Turner. London: Verso, 2001.

——. *The Perfect Crime*. Translated by Chris Turner. London: Verso, 1996.

——. "The Precession of Simulacra." In *Simulacra and Simulation*, 1-42.

——. "Simulacra and Science Fiction." In *Simulacra and Simulation*, 121-27.

——. *Simulacra and Simulation*. Translated by Sheila Faria Glaser. Ann Arbor: University of Michigan Press, 1994.

——. *Symbolic Exchange and Death*. Translated by Iain Hamilton Grant. London: Sage Publications, 1993.

Baxter, Stephen. "The Logic Pool." In *Nanotech,* edited by Jack Dann and Gardner Dozois, 153-73. New York: Ace Books, 1998.

Bear, Greg. *Blood Music*. New York: Arbor House, 1985. Reprint, New York: I Books, 2002.

——. *Queen of Angels*. New York: Warner Books, 1990.

——. *Slant*. New York: Tor, 1991.

Beck, Ulrich. *Risk Society: Towards a New Modernity.* Translated by Mark Ritter. London: Sage Publications, 1992.

Benford, Gregory. "Bio/Nano/Tech." In *Nanodreams,* edited by Elton Elliott, 192-200. New York: Baen Books, 1995.

Benjamin, Walter. "Theses on the Philosophy of History." In *Illuminations,* 253-64. Translated by Harry Zohn. New York: Shocken Books, 1968. "A Obra De Arte Na Época De Sua Reprodutibilidade Técnica." In *Teoria Da Cultura De Massa,* edited by Luiz Costa Lima, 205-40. São Paulo: Paz e Terra, 1990.

Benveniste, Emile. *Problems in General Linguistics.* Translated by Mary Elizabeth Meek. Coral Gables, FL: University of Miami Press, 1971.

Bergvall, Caroline. *Ambient Fish.* <http://epc.buffalo.edu/authors/bergvall/amfish/cbflash.html>.

———. *Goan Atom.* San Francisco: Krupskaya, 2001.

Blish, James. "Surface Tension." In *The Science Fiction Hall of Fame: Volume I: The Greatest Science Fiction Stories of All Time,* edited by Robert Silverberg, 477-514. New York: Avon Books, 1971.

Block, Steven M. "What Is Nanotechnology?" Keynote presentation at National Institutes of Health conference, "Nanoscience and Nanotechnology: Shaping Biomedical Research", Natcher Conference Center, Bethesda, Maryland, June 25, 2000.

Bök, Christian. *Crystallography.* 1994. 2nd ed. Toronto: Coach House, 2003.

———. *'Pataphysics: The Poetics of an Imaginary Science.* Evanston, IL: Northwestern University Press, 2002.

Borges, Jorge Luis. "The Garden of Forking Paths." In *Labyrinths: Selected Stories & Other Writings,* edited by Donald A. Yates and James E. Irby, 19-28. New York: New Directions, 1962.

Brown, Charles N., and William G. Contento. "The Locus Index to Science Fiction (1984-1998)." <http://www.locusmag.com/index/0start.htm.#TOC> (accessed October 17, 2003).

Buiani, Roberta. "Virtual Museums and the Web: A Dilemma of Compatibility?" In *Proceedings of Life by Design: Everyday Digital Culture Conference,* 6-17. University of California, Irvine, April 10-12, 2003.

Bukatman, Scott. *Terminal Identity: The Virtual Subject in Postmodern Science Fiction.* Durham, N.C.: Duke University Press, 1993.

———. "There's Always Tomorrowland: Disney and the Hypercinematic Experience." *October* 57 (1991): 55-78.

Butler, Octavia. *Parable of the Sower.* New York: Four Walls Eight Windows, 1993.

Calvino, Italo. *Invisible Cities.* New York: Harcourt Brace Jovanovich, 1974.

Cameron, James and Charles H. Eglee. "Episode 19: Hit A Sista Back." Disc 6. *Dark Angel,* collector's ed. DVD. Directed by James Whitmore. 2001; Beverly Hills, CA: Twentieth Century Fox, 2003.

Cameron, Susan. *Choosing Not Choosing: Dickinson's Fascicles.* Chicago: Chicago University Press, 1992.

Carroll, Lewis, and Donald J. Gray. *Alice in Wonderland: A Norton Critical Edition.* 2nd ed. New York: W.W. Norton & Co, 1992.

Castells, Manuel. *End of Millennium.* Part III of *The Information Age: Economy, Society and Culture.* Oxford: Blackwell Publishers, 1998.

———.*The Rise of the Network Society.* Malden, MA: Blackwell Publishers, 2000.

Cayley, John. "The Code Is Not the Text (Unless It Is the Text)." *Electronic Book Review* (September 10, 2002). <http://www.electronicbookreview.com/v3/servlet/ebr?command=view_essay&essay_id=cayleyele> (accessed October 15, 2003).

Chakraborty, Tapash. *Quantum Dots.* New York: Elsevier Health Sciences, 1999.

Chaos Computer Club. *Blinkenlights.* <http://www.blinkenlights.de/> (accessed May 25, 2003).

Chen, Julian C. *Introduction to Scanning Tunneling Microscopy.* New York: Oxford University Press, 1993.

Clareson, Thomas D. *Some Kind of Paradise: The Emergence of American Science Fiction.* Westport, CT: Greenwood Press, 1985.

Clarke, Arthur C. *Profiles of the Future: An Inquiry into the Limits of the Possible.* New York: Harper & Row, 1958.

Clinton, William J. Address to Caltech on Science and Technology. California Institute of Technology, January 21, 2000. Videorecording of talk produced by Caltech's Audio Visual Services, Electronic Media Publications, and Digital Media Center, made available at <pr.caltech.edu/events/presidential_speech//PresVisit-MCP-LAN.ram>.

Colbert, Daniel T., and Richard E. Smalley. "Fullerine Nanotubes for Molecular Electronics." *Trends in Biotechnology* 17 (1999): 46-50.

Coleridge, Samuel Taylor. "Biographia Literaria." 1817.

Collins, Graham P. "Shamans of Small." *Scientific American* 285, no. 3 (2001): 86-91.

Connected Cities – Kunstprozesse im urbanen Netz. <http://www.connected-cities.de/> (accessed June 20, 2003).

Cramer, John G. "Nanotechnology: The Coming Storm." Foreword to *Nanodreams,* edited by Elton Elliott, 4-12. New York: Baen Books, 1995.

Crandall, B.C., ed. *Nanotechnology: Molecular Speculations on Global Abundance.* Cambridge: MIT Press, 1996.

Crichton, Michael. *Prey: A Novel.* New York: Harper Collins, 2002.

Crommie, M.F., C.P. Lutz, and D.M. Eigler. "Imaging Standing Waves in a Two Dimensional Electron Gas." *Nature* 363 (1993): 524-27.

Cuberes, M.T., R.R. Schlittler, and J.K. Gimzewski. "Room-Temperature Repositioning of Individual C_{60} Molecules at Cu Steps: Operation of a Molecular Counting Device." *Applied Physics Letters* 69, no. 20 (1996): 3016-18.

Culler, Jonathan. *The Pursuit of Signs: Semiotics, Literature, Deconstruction.* London: Routledge & Kegan Paul, 1981.

Cummings, Ray. "The Girl in the Golden Atom." In *Under the Moons of Mars: A History and Anthology of "the Scientific Romance" in the Munsey Magazines, 1912-1920,* edited by Sam Moskowitz, 175-220. New York: Holt, Rinehart and Winston, 1970.

Dai, Hongjie, Nathan Franklin, and Jie Han. "Exploiting the Properties of Carbon Nanotubes for Nanolithography." *Applied Physics Letters* 73 (1998): 1508-10.

Daniel, Tony. *Metaplanetary: A Novel of Interplanetary Civil War.* New York: EOS, 2001.

Dann, Jack, and Gardner Dozois, eds. *Nanotech.* New York: Ace Books, 1998.

Davis, Anthony P. "Synthetic Molecular Motors." *Nature* 401 (1999): 120-21.

Deleuze, Gilles. "Ideas and the Synthesis of Difference." In *Difference and Repetition,* 168-221. New York: Columbia University Press, 1994.

DeLillo, Don. *Cosmopolis: A Novel.* New York: Scribner, 2003.

———. *White Noise.* Edited by Harold Bloom. Philadelphia: Chelsea House Publishers, 2003.

Derrida, Jacques. "The Ends of Man." In *Margins of Philosophy,* 109-36. Translated by Alan Bass. Chicago: University of Chicago Press, 1982.

———. *Of Grammatology.* Translated by Gayatri Spivak. Baltimore: Johns Hopkins University Press, 1974.

——. "Structure, Sign and Play in the Discourse of the Human Sciences." In *Writing and Difference*, 278-93. Translated by Alan Bass. Chicago: University of Chicago Press, 1978.

Di Filippo, Paul. "Any Major Dude." In *Nanotech*, edited by Jack Dann and Gardner Dozois, 174-202. New York: Ace Books, 1998.

——. *Ribofunk*. New York: Four Wall Eight Windows, 1996.

Dickinson, Emily. *The Complete Poems of Emily Dickinson*. Edited by Thomas H. Johnson. Boston: Little, Brown and Co., 1960.

——. *The Manuscript Books of Emily Dickinson*. Vol. 2. Edited by R. W. Franklin. Cambridge: Belknap Press, 1981.

Dozois, Gardner. Moderator. "Nanotechnology - with M. Stephen Gillett, Todd Washington, Kathleen Ann Goonan, and Wil McCarthy." SciFi.com Chat. February 22, 2000. <http://www.scifi.com/transcripts/2000/nanotech.html>.

Drexler, Eric K. *Engines of Creation: The Coming Era of Nanotechnology*. New York: Anchor Books, 1986.

——. "From Nanodreams to Reality." Introduction to *Nanodreams*, edited by Elton Elliott, 13-16. New York: Baen Books, 1995.

——. "Machine-Phase Nanotechnology." In Fritz, comp., *Understanding Nanotechnology*, (2002): 104-08.

——. "Machines of Inner Space." In *Nanotechnology Research and Perspectives: Papers from the First Foresight Conference on Nanotechnology*, edited by B.C. Crandall and James Lewis. Cambridge: MIT Press, 1992.

——. "Molecular Engineering: An Approach to the Development of General Capabilities for Molecular Manipulation." *Proceedings of the National Academy of Sciences* 78 (1981): 5275-78.

——. "Molecular Manufacturing as a Path to Space." In *Prospects in Nanotechnology*, edited by Marcus Krummenacker and James Lewis, 197-205. New York: Wiley, 1995.

——. *Nanosystems: Molecular Machinery, Manufacturing, and Computation*. New York: John Wiley & Sons, 1992.

Drexler, K. Eric, Chris Peterson, and Gayle Pergamit. *Unbounding the Future: The Nanotechnology Revolution*. New York: Quill, 1992.

Du Charme, Wesley M. *Becoming Immortal: Nanotechnology, You, and the Demise of Death*. Evergreen, CO: Blue Creek Ventures, 1995.

Dworkin, Craig. *Reading the Illegible*. Evanston, IL: Northwestern University Press, 2003.

Egan, Greg. "Axiomatic." In *Nanotech*, edited by Jack Dann and Gardner Dozois, 40-57. New York: Ace Books, 1998.

——. "Dust." In *The Year's Best Science Fiction: Tenth Annual Collection*, edited by Gardner Dozois, 87-112. New York: St. Martin's, 1993.

——. *Permutation City*. New York: Harper Prism, 1995.

——. "Reasons to Be Cheerful." In *The Year's Best Science Fiction: Fifteenth Annual Collection*, edited by Gardner Dozois, 69-94. New York: St. Martin's, 1998.

Eigler, D.M. "From the Bottom Up: Building Things with Atoms." In *Nanotechnology*, edited by Gregory Timp, 425-35. New York: Springer-Verlag, 1999.

Eigler, D.M., C.P. Lutz, and W.E. Rudge. "An Atomic Switch Realized with the Scanning Tunneling Microscope." *Nature* 352 (1991): 600-02.

Eigler, D.M., and E.K. Schweizer. "Positioning Single Atoms with a Scanning Tunneling Microscope." *Nature* 344 (1990): 524-26.

Elliott, Elton. *Nanodreams*. New York: Baen Books, 1995.

Engler, Craig E. "It Takes a Rocket Scientist: Amazon.com Interviews Wil McCarthy." Amazon.com 1999. <http://www.cyberhaven.com/books/sciencefiction/mccarthy.html>.

Epstein, Edward. "Silicon Valley Pins Hopes on Nanotechnology Boom: U.S. Ready to Spend Billions on Revolutionary Science." *San Francisco Chronicle,* May 8 2003. <http://sfgate.com/cgi-bin/ article.cgi?f=/c/a/2003/05/08/MN286866.DTL>.

Farr, Judith. *The Passion of Emily Dickinson*. Cambridge: Harvard University Press, 1992.

Feder, Barnaby J. "In the World of the Very Small, Companies Make Big Plans." *New York Times,* sec. C, p. 19, 16 December 2002.

Feynman, Richard. "There's Plenty of Room at the Bottom." Speech given at the Annual Meeting of the American Physical Society, California Institute of Technology, Pasadena, CA, December 29, 1959. Originally published in *Engineering and Science* 23, no. 5 (February 1960): 22-36. Reprinted in *Miniaturization,* edited by H.D. Gilbert, 282-96. New York: Reinhold, 1961. Reprinted in *Nanotechnology Research and Perspectives: Papers from the First Foresight Conference on Nanotechnology,* ed. B.C. Crandall and James Lewis (Cambridge, Mass.: MIT Press, 1992). Available online at <http://www.zyvex.com/nanotech/feynman.html> (accessed October 20, 2003).

Flynn, Michael F. *The Nanotech Chronicles*. New York: Baen Books, 1991.

———. "Remember'd Kisses." In *Nanotech,* edited by Jack Dann and Gardner Dozois, 58-98. New York: Ace Books, 1998.

Foresight Institute. <http://www.foresight.org> (accessed 2003).

Forsythe, Diana E. *Studying the People Who Study Us: An Anthropologist in the World of Artificial Intelligence*. Stanford: Stanford University Press, 2001.

Foster, J.S., J.E. Frommer, and P.C. Arnett. "Molecular Manipulation Using a Tunneling Microscope." *Nature* 331 (1988): 324-26.

Foucault, Michel. "Of Other Spaces (1967), Heterotopias." Translated by Jay Miskowiec. <http: //foucault.info/documents/foucault.heteroTopia.en.html> (accessed October 16, 2003). Originally published as "Des Espace Autres." *Architecture /Mouvement/ Continuité* (October 1984).

———. Foucault, Michel. *The Order of Things: An Archaeology of the Human Sciences*. New York: Vintage Books, 1973.

Freitas, Robert A., Jr. *Nanomedicine*. Vol. 1: *Basic Capabilities*. Georgetown, TX: Landes Bioscience, 1999.

Fritz, Sandy, comp. *Understanding Nanotechnology: From the Editors of Scientific American*. New York: Warner Books, 2002.

Gamow, George. *Mr. Tompkins in Wonderland*. Cambridge: Cambridge University Press, 1940.

Gibson, William. *Mona Lisa Overdrive*. New York: Bantam Books, 1989.

———. *Neuromancer*. New York: Ace Books, 2000.

———. *Pattern Recognition*. New York: G.P. Putnam's Sons, 2003.

Giddens, Anthony. *The Consequences of Modernity*. Stanford: Stanford University Press, 1990.

———. *Runaway World: How Globalization Is Reshaping Our Lives*. London: Profile Books, 2002.

Gimzewski, James. Interview by Rebecca N. Lawrence. *BioMedNet* 118 (January 18, 2002). <http: //news.bmn.com/hmsbeagle/118/notes/biofeed> (accessed May 30, 2003).

Gimzewski, James, and Victoria Vesna. "The Nanomeme Syndrome: Blurring of Fact and Fiction in the Construction of a New Science." *Technoetic Arts* Journal, 1, no. 1 (May 2003): 7-24. <http: //vv.arts.ucla.edu/publications/ publications_frameset.htm> (accessed October 20, 2003).

Gimzewski, James K., and Christian Joachim. "Nanoscale Science of Single Molecules Using Local Probes." *Science* 283 (1999): 1683-88.

Gitelman, Lisa. *Scripts, Grooves, and Writing Machines: Representing Technology in the Edson Era.* Stanford: Stanford University Press, 1999.

Goonan, Kathleen Ann. "Chicon Live Chat." 27 May 2002. <http://www.cybling.com/chicon/guests/Goonan_Kathleen.html>.

———. *Crescent City Rhapsody.* New York: EOS, 2000.

———. "Extending Our Senses." *Locus* 46, no. 6 (2001): 8, 83-84.

———. *Light Music.* New York: EOS, 2002.

———. *Mississippi Blues.* New York: Tor Books, 1997.

———. "Nanotechnology II: Kathleen Ann Goonan and Wil McCarthy." SciFi.com Chat. May 4, 2000. <http://www.scifi.com/transcripts/2000/nanotech2.html>.

———. *Queen City Jazz.* New York: Tor Books, 1996.

———. "Sunflowers." In *Nanotech,* edited by Jack Dann and Gardner Dozois, 117-52. New York: Ace Books, 1998.

Gray, Chris Hables. *Cyborg Citizen: Politics in the Posthuman Age.* New York: Routledge, 2001.

Grogaard, Stian. "Low Tech-High Concept: Digital Media, Art, and the State of the Arts." In *Digital Media Revisited,* edited by Gunnar Liestol, Andrew Morrison and Terje Rasmussen. Cambridge: MIT Press, 2003.

Halberstam, Judith, and Ira Livingston, eds. *Posthuman Bodies.* Bloomington: Indiana University Press, 1995.

Hall, J. Storrs. "An Overview of Nanotechnology." Sci.nanotech Info Online (1995). <http://nanotech.dyndns.org/sci.nanotech/overview.html> (accessed 2003).

———. "Utility Fog: The Stuff That Dreams Are Made Of." In *Nanotechnology: Molecular Speculations on Global Abundance,* edited by B.C. Crandall, 161-84. Cambridge: MIT Press, 1996.

Hansen, Mark. *Embodying Technesis: Technology Beyond Writing.* Ann Arbor: Michigan University Press, 2000.

Haraway, Donna J. "The Biopolitics of Postmodern Bodies: Constitutions of Self in Immune System Discourse." In *Simians, Cyborgs, and Women: The Reinvention of Nature,* 203-30. New York: Routledge, 1991.

———. "A Cyborg Manifesto: Science, Technology, and Socialist-Feminism in the Late Twentieth Century." In *Simians, Cyborgs, and Women: The Reinvention of Nature,* 149-81. New York: Routledge, 1991.

Hartman, Geoffrey. *Criticism in the Wilderness: The Study of Literature Today.* New Haven: Yale University Press, 1980.

Hartwell, David G., and Kathryn Cramer, eds. *The Ascent of Wonder: The Evolution of Hard Sf.* New York: Tor Books, 1994.

Hawkins, Gary. "Constructing and Residing in the Paradox of Dickinson's Prismatic Space." *The Emily Dickinson Journal* 9, no. 1 (2000): 49-70.

Hayles, N. Katherine. *The Cosmic Web: Scientific Field Models and Literary Strategies in the Twentieth Century.* Ithaca: Cornell University Press, 1984.

———. "Flesh and Metal: Reconfiguring the Mindbody in Virtual Environments." *Configurations* 10 (2002): 297-320.

———. *How We Became Posthuman: Virtual Bodies in Cybernetics, Literature, and Informatics.* Chicago: University of Chicago Press, 1999.

——. "Translating Media: Why We Should Rethink Textuality." *The Yale Journal of Criticism* (forthcoming, fall 2003).

——. *Writing Machines.* Cambridge: MIT Press, 2002.

Hegel, G.F.W. *Phenomenology of Spirit.* Translated by A.V. Miller. Oxford: Oxford University Press, 1977.

Heidegger, Martin. "Poetically Man Dwells." In *Poetry, Language, Thought.* Translated by Albert Hofstadter. New York: HarperCollins, 1971.

——. "The Question Concerning Technology." In *Basic Writings.* Translated by William Lovitt. New York: HarperCollins, 1993.

Heinlein, Robert A. "Waldo." In *Waldo & Magic, Inc.,* 1-154. New York: Dell Rey, 1986.

Houellebecq, Michel. *The Elementary Particles.* Translated by Frank Wynne. New York: Vintage Books, 2000.

Howe, Susan. *My Emily Dickinson.* Berkeley: North Atlantic, 1985.

Hunt, Peter. *Children's Literature: An Anthology, 1801-1902.* Malden: Blackwell, 2001.

Huntington, John. *Rationalizing Genius: Ideological Strategies in the Classic American Science Fiction Short Story.* New Brunswick: Rutgers Univ Press, 1989.

Hurley, Kelly. "Reading Like an Alien: Posthuman Identity in Ridley Scott's Alien and David Cronenberg's Rabid." In *Posthuman Bodies,* edited by Judith Halberstam and Ira Livingston. Bloomington: Indiana University Press, 1995.

IBM Research Website. <http://www.almaden.ibm.com/vis/stm/atomo.html#stm10>.

Institute für Allgemeine Physik. *Slideshow.* <http://www.iap.tuwien.ac.at/www/surface/STM_Gallery/ stm_schematic.html>.

International Technology Research Institute, World Technology (WTEC) Division. "Nanotechnology Database." <http://www.wtec.org/loyola/nano/links.htm> (accessed 2002).

Jakobsen, Roman. *Selected Writings.* Vol. 2. The Hague: Mouton, 1971.

Jameson, Frederic. "Future City." *New Left Review* 21 (May/June 2003): 65-79.

——. *Postmodernism: Or, the Cultural Logic of Late Capitalism.* Durham, NC: Duke University Press, 1991.

Jones, David E. H. "Technical Boundless Optimism." *Nature* 374 (1995): 835-37.

Jones, Jo Elwyn, and J. Francis Gladstone. *The Alice Companion: A Guide to Lewis Carroll's Alice Books.* Houndmills, Basingstoke, Hampshire and London: Macmillan Press LTD, 1998.

Jones, Steven Swann. *The Fairy Tale: The Magic Mirror of Imagination.* New York: Routledge, 2002.

Joy, Bill. "Why the Future Doesn't Need Us." *Wired* 8, no. 4 (April 2000): 238-63. <http://www.wired.com/ wired/archive/8.04/joy.html> (accessed October 20, 2003).

Joyce, Michael. *Of Two Minds: Hypertext Pedagogy and Poetics.* Ann Arbor: University of Michigan Press, 1995.

Juhasz, Suzanne. *The Undiscovered Continent: Emily Dickinson and the Space of the Mind.* Bloomington: Indiana University Press, 1983.

Kher, Inder Nath. *The Landscape of Absence.* New Haven: Yale University Press, 1974.

Kim, Philip, and Charles M. Lieber. "Nanotube Nanotweezers." *Science* 286 (1999): 2148-50.

Kittler, Friedrich. *Discourse Networks, 1800/1900.* Translated by Michael Metteer with Chris Cullens. Palo Alto: Stanford University Press, 1990.

——. *Gramophone, Film, Typewriter.* 1986. Translated by Geoffrey Winthrop-Young and Michael Wurtz. Palo Alto: Stanford University Press, 1999.

——. "There Is No Software." *Ctheory.net* (October 18, 1995). <http://www.ctheory.net/text_file.asp?pick=74> (accessed October 15, 2003).

Klein, Julie Thompson. *Crossing Boundaries: Knowledge, Disciplinarities, and Interdisciplinarities.* Charlottesville: University Press of Virginia, 1996.

Klein, Julie Thompson *et al. Transdisciplinarity: Joint Problem-Solving among Science, Technology, and Society,* eds. Julie Thompson Klein, et. al. (Basel: Birkhäuser Verlag, 2001).

Knight, Jonathan. "The Engine of Creation." *New Scientist* 162, no. 2191 (1999): 38-41.

Kohn, Denise. "'I Cannot Live with You -'." In *An Emily Dickinson Encyclopedia,* edited by Jane Donahue Eberwein, 154-5. Westport, CT: Greenwood Press, 1998.

Koolhaas, Rem. "Junkspace." *October* 100 (Spring, 2002): 175-90.

Kress, Nancy. "Margin of Error." In *Nanotech,* edited by Jack Dann and Gardner Dozois, 32-39. New York: Ace Books, 1998.

———. "The Most Famous Little Girl in the World." In *The Year's Best Science Fiction: Twentieth Annual Collection,* edited by Gardner Dozois, 52-70. New York: St. Martin's Griffin, 2003.

Kristeva, Julia. *Revolution in Poetic Language.* Translated by Margaret Waller. New York: Columbia University Press, 1984.

Krummenacker, Markus. "Steps Towards Molecular Manufacturing." *Chemical Design Automation News* 9 (1994): 1, 29-39.

Krummenacker, Markus, and James Lewis, eds. *Prospects in Nanotechnology: Toward Molecular Manufacturing.* New York: Wiley, 1995.

Landis, Geoffrey A. "Willy in the Nano-Lab." In *Nanotech,* edited by Jack Dann and Gardner Dozois, 274. New York: Ace Books, 1998.

Landon, Brooks. *Science Fiction after 1900: From the Steam Man to the Stars.* New York: Routledge, 2002.

Latour, Bruno. "Research Space: The World Wide Lab." *Wired* (June, 2003).

———. *We Have Never Been Modern.* Translated by Catherine Porter. Cambridge: Harvard University Press, 1993.

Laurel, Brenda. *Computers as Theater.* Reading: Addison-Wesley Publishing Company, 1993.

LeGuin, Ursula K. *The Left Hand of Darkness.* 1969. New York: Harper, 1976.

Lenoir, Timothy. *Instituting Science: The Cultural Production of Scientific Disciplines.* Stanford: Stanford University Press, 1997.

———. "Introduction." In *Inscribing Science: Scientific Texts and the Materiality of Communication.* Stanford: Stanford University Press, 1998.

Levinas, Emmanuel. *Otherwise Than Being, or Beyond Essence.* Translated by Alphonso Lingis. Pittsburgh: Duquesne University Press, 1981.

———. *Totality and Infinity: An Essay on Exteriority.* Translated by Alphonso Lingis. Pittsburgh: Duquesne University Press, 1981.

Levy, Pierre. *Cyberculture.* Translated by Robert Bononno. Minneapolis: University of Minnesota Press, 1997.

Lieber, Charles. "Scanning Tunneling Microscopy." *Chemical & Engineering News* 18 Apr. 1994, 28-43.

Little, T.E. *The Fantasts: Studies in J.R.R. Tolkien, Lewis Carroll, Mervyn Peake, Nikolay Gogol, and Kenneth Grahame.* Amersham, England: Avebury, 1984.

Loyer, Erik. *Chroma.* <http://www.marrowmonkey.com/chroma.html> (accessed October 15, 2003).

Lozano-Hemmer, Rafael. *Vectorial Elevation.* <http://www.fundacion.telefonica.com/at/rlh/> (accessed May 25, 2003).

Luhmann, Niklas. *Observations on Modernity.* Translated by William Whobrey. Stanford: Stanford University Press, 1998.

———. *The Reality of the Mass Media.* Translated by Kathleen Cross. Stanford: Stanford University Press, 2000.

Lunenfeld, Peter. "Screen Grabs: The Digital Dialectic and New Media Theory." In *The Digital Dialectic: New Essays on New Media.* Cambridge: MIT Press, 1999.

———. *Snap to Grid: A User's Guide to Digital Arts, Media, and Cultures.* Cambridge: MIT Press, 2000.

———. "Unfinished Business." In *The Digital Dialectic: New Essays on New Media.* Cambridge: MIT Press, 1999.

MacLeod, Ian R. "New Light on the Drake Equation." In *The Year's Best Science Fiction: Nineteenth Annual Collection,* edited by Gardner Dozois, 1-43. New York: St. Martin's Griffin, 2002.

Manlove, Colin N. "Victorian and Modern Fantasy: Some Contrasts." In *The Celebration of the Fantastic: Selected Papers from the Tenth Anniversary International Conference on the Fantastic in the Arts,* edited by Csilla Bertha, Marshall B. Tymn and Donald E. Morse, 9-22. Westport: Greenwood Press, 1992.

Manovich, Lev. *The Language of New Media.* Cambridge: MIT Press, 2001.

———. "The Poetics of Augmented Space: Learning from Prada." <http://manovich.net/DOCS/augmented_space.doc> (accessed May 25, 2003).

Marusek, David. "We Were out of Our Minds with Joy." In *Nanotech,* edited by Jack Dann and Gardner Dozois, 203-73. New York: Ace Books, 1998.

Massumi, Brian. *Parables for the Virtual: Movement, Affect, Sensation.* Durham, NC: Duke University Press, 2002.

McCaffery, Steve. *Carnival: The Second Panel, 1970-1975.* <http://www.chbooks.com/online/carnival/index.html>.

———. "Knowledge Never Knew." In *Seven Pages Missing: Volume One.* Toronto: Coach House Press, 2000.

———. *Prior to Meaning: The Protosemantic and Poetics.* Evanston, IL: Northwestern University Press, 2001.

———. *Seven Pages Missing: Volume One.* Toronto: Coach House Press, 2000.

———. "Shifters." In *Seven Pages Missing: Volume One.* Toronto: Coach House Press, 2000.

———. "Shifters: A Note." In *Seven Pages Missing: Volume One.* Toronto: Coach House Press, 2000.

McCarthy, Wil. *Bloom.* New York: Del Rey Books, 1998.

———. *The Collapsium.* New York: Ballantine/Del Rey Books, 2000.

McCarthy, Wil. *Hacking Matter: Levitating Chairs, Quantum Mirages, and the Infinite Weirdness of Programmable Atoms.* New York: Basic Books, 2003.

———. "Nanotechnology: Abuses of, and Replacements For." *The Bulletin of the Science Fiction and Fantasy Writers of America* 151 (2001): 20-23.

———. *The Wellstone.* New York: Bantam Spectra, 2003.

McDonald, Ian. *Evolution's Shore.* (British title: *Chaga).* New York: Bantam Spectra, 1995.

———. "Recording Angel." In *Nanotech,* edited by Jack Dann and Gardner Dozois, 99-116. New York: Ace Books, 1998.

———. "Tendeleo's Story." In *The Year's Best Science Fiction: Eighteenth Annual Collection,* edited by Gardner Dozois, 558-609. New York: St. Martin's Griffin, 2001.

McHale, Brian. *Postmodernist Fiction.* New York: Routledge, 1997.

McKendree, Tom. "Nanotech Hobbies." In *Nanotechnology,* edited by B.C. Crandall, 135-44. Cambridge, MA: MIT Press, 1996.

McKibben, Bill. *Enough: Staying Human in an Engineered Age.* New York: Times Books, 2003.

McLuhan, Marshall. *Understanding Media: The Extensions of Man.* New York: McGraw-Hill, 1964.

Merkle, Ralph C. "Cryonics." <http://www.merkle.com/cryo/> (accessed 2003).

———. "Design-Ahead for Nanotechnology." In *Prospects in Nanotechnology,* edited by Marcus Krummenacker and James Lewis, 23-52. New York: Wiley, 1995.

———. "It's a Small, Small, Small, Small World." *Technology Review* 100, no. 2 (1997): 25-32.

———. "Letter to the Editor." *Technology Review* 102, no. 3 (1999): 15-16.

———. "Molecular Manufacturing: Adding Positional Control to Chemical Synthesis." *Chemical Design Automation News* 8 (1993): 1, 55-61.

———. "Nanotechnology and Medicine." In *Advances in Anti-Aging Medicine,* vol. 1, edited by Ronald M. Klatz and Francis A. Kovarik, 277-86. Larchmont, NY: Liebert, 1996.

———. "A Response to Scientific American's News Story Trends in Nanotechnology." Foresight Institute. 1996. <http://www.islandone.org/Foresight/SciAmDebate/SciAmResponse.html>.

———. "Self-Replicating Systems and Molecular Manufacturing." *Journal of the British Interplanetary Society* 45 (1992): 407-13.

Merlau-Ponty, Maurice. *The Visible and Invisible.* Translated by Alphonso Lingis. Edited by Claude Lefort. Evanston, IL: Northwestern University Press, 1968.

Miller, Christanne. *Emily Dickinson: A Poet's Grammar.* Cambridge: Harvard University Press.

Milne, Drew. "A Veritable Dollmine." *Jacket Magazine* (July 2000). <http://jacketmagazine.com/12/milne-bergvall.html>.

Mirkin, Chad A. "Tweezers for the Nanotool Kit." *Science* 286 (1999): 2095-96.

Mitchell, William J. *City of Bits. Space, Place and the Infobahn.* Cambridge: MIT Press, 1999.

Mitchell, W.J.T., ed. *Language of Images.* Chicago: University of Chicago Press, 1974.

Moravec, Hans. *Mind Children: The Future of Robot and Human Intelligence.* Cambridge: Harvard University Press, 1988.

More, Max. "The Extropian Principles 3.0." 1998. Extropy Institute. <http://www.extropy.com/ideas/principles.html>.

Museo Virtual de Artes El País (MUVA).© 1997 EL PAIS. <http://www.elpais.com.uy/muva2/> (accessed September 30, 2003).

Myers, Greg. "Scientific Speculation and Literary Style in a Molecular Genetics Article." *Science in Context* 4 (1991): 321-46.

Nagata, Linda. *The Bohr Maker.* New York: Bantam Spectra, 1995.

———. *Deception Well.* New York: Bantam Spectra, 1997.

———. *Vast.* New York: Bantam Spectra, 1998.

NanoInvestor News Website. <http://www.nanoinvestornews.com> (accessed October 20, 2003).

Napier, Anthony S. *The Nanotechnology in Science Fiction Bibliography.* Issue 29, December 5, 2002. <http://www.geocities.com/asnapier/nano/n-sf/>.

National Nanofabrication Users Network. <http://www.nnun.org> (accessed 2003).

National Nanotechnology Initiative. <http://www.nano.gov> (accessed 2003).

National Science and Technology Council. *National Nanotechnology Initiative: Leading the Next Industrial Revolution.* Washington, D.C.: Office of Science and Technology Policy, 2000.

National Science Foundation at University of California (UC). "Exploring the World at Nanoscale, summer 2001. <http://www.ee.ucr.edu/~alexb/uc-nano.pdf>.

National Science Foundation. "NSF Awards New Grants to Study Societal Implications of Nanotechnology." NSF PR 03-89, August 25, 2003.

———. "Societal Implications of Nanoscience and Nanotechnology." Final Report from the Workshop held at the National Science Foundation (Sept. 28-29, 2000). March 2001.

Negroponte, Nicholas. *Being Digital.* New York: Alfred A. Knopf, 1996.

Nicholls, Peter. "Big Dumb Objects." In *The Encyclopedia of Science Fiction,* edited by John Clute and Peter Nicholls, 118-19. New York: St. Martin's Press, 1993.

Nye, David E. *American Technological Sublime.* Cambridge: MIT Press, 1994.

———. Nye, David. *Narratives and Spaces: Technology and the Construction of American Culture.* Exeter: University of Exeter Press, 1997.

Ohta, Yuichi, and Hideyuki Tamura. *Mixed Reality: Merging Real and Virtual Worlds.* Tokyo: Ohmsha, 1999.

Oppen, George. "Route." In *George Oppen: New Collected Poems,* edited by Michael Davidson. New York: New Directions, 2002.

Ostman, Charles. "The Creative and Design Intelligence Evolution in Today's Technological Innovation Roundtable." Presentation at the "Visionary Forum 2003: Nanotechnology and Beyond," Jet Propulsion Laboratory, La Cañada Flintridge, CA, 22 July 2003.

Otis, Laura. *Membranes: Metaphors of Invasion in Nineteenth-Century Literature, Science, and Politics.* Baltimore: Johns Hopkins University Press, 2000.

Padgett, Lewis. "Mimsy Were the Borogoves." In *The Science Fiction Hall of Fame: Volume I: The Greatest Science Fiction Stories of All Time,* edited by Robert Silverberg, 226-60. New York: Avon Books, 1970.

Panshin, Alexi, and Cory Panshin. *The World Beyond the Hill: Science Fiction and the Quest for Transcendence.* Los Angeles: Jeremy P. Tarcher, 1989.

Parente, Andre. *O Virtual E O Hipertextual.* Rio de Janeiro: Pazulin, 1999.

Petersen, Aage. "The Philosophy of Neils Bohr." *Bulletin of the Atomic Scientist* 19 (September 1963): 10.

Peterson, Christine L. "Nanotechnology: Evolution of the Concept." In *Prospects in Nanotechnology,* edited by Marcus Krummenacker and James Lewis, 173-86. New York: Wiley, 1995.

Phoenix, Chris. "Don't Let Crichton's *Prey* Scare You—the Science Isn't Real." *Nanotechnology Now.* <http://nanotech-now.com> (accessed January 2003).

Postman, Neil. *Technopoly: The Surrender of Culture to Technology.* New York: Vintage, 1992.

Pound, Ezra. *Guide to Kulchur.* New York: New Directions, 1938.

Pynchon, Thomas. *Gravity's Rainbow.* 1973. New York: Penguin, 2000.

Radford, Tim. "Brave New World or Miniature Menace? Why Charles Fears Grey Goo Nightmare: Royal Society Asked to Look at Risks of Nanotechnology." *The Guardian,* 29 April 2003.

Raley, Rita. "Reveal Codes: Hypertext and Performance." *Postmodern Culture* 12, no. 1 (2001). <http://muse.jhu.edu/journals/pmc/v012/12.1raley.html> (accessed October 15, 2003).

Rapport, David and Margaret A. Somerville. *Transdisciplinarity: Recreating Integrated Knowledge.* Montréal: McGill-Queen's University Press, 2003.

Rasula, Jed. "The Poetics of Embodiment: A Theory of Exceptions." Ph.D. diss., University of California at Santa Cruz, 1989.

Ratner, Mark, and Daniel Ratner. *Nanotechnology: A Gentle Introduction to the Next Big Idea.* Upper Saddle River, NJ: Prentice Hall, 2003.

Rawstern, Rocky. "Interview with Author Wil McCarthy—June 2003." Nanotechnology Now. 13 June 2003. <http://nanotech-now.com/Wil-McCarthy-interview-06132003.htm>.

Reed, Mark. "Quantum Constructions." *Science* 262 (1993): 195.

Regis, Ed. *Great Mambo Chicken and the Transhuman Condition: Science Slightly over the Edge.* New York: Addison-Wesley, 1990.

———. *Nano. The Emerging Science of Nanotechnology: Remaking the World-Molecule by Molecule.* Boston: Little, Brown and Co., 1995.

Reynolds, Alastair. "Glacial." In *The Year's Best Science Fiction: Nineteenth Annual Collection,* edited by Gardner Dozois, 247-80. New York: St. Martin's Griffin, 2002.

———. "Great Wall of Mars." In *The Year's Best Science Fiction: Eighteenth Annual Collection,* edited by Gardner Dozois, *307-39.* New York: St. Martin's Griffin, 2001.

"Richard Feynman." Photosynthesis.Sound.com. 2002. <http://www.photosynthesis.com/Richard_Feynman.html>.

Rotman, David. "Will the Real Nanotech Please Stand Up?" *Technology Review* 102, no. 2 (1999): 47-53.

Roukes, Michael L. "Forward." In Fritz, comp., *Understanding Nanotechnology,* vii-x.

———. "Plenty of Room, Indeed." In Fritz, comp., *Understanding Nanotechnology,* 56-71.

Ryan, Marie-Laure. *Narrative as Virtual Reality: Immersion and Interactivity in Literature and Electronic Media.* Baltimore: Johns Hopkins University Press, 2001.

Ryman, Geoff. "Afterword." In *Unconquered Countries,* 273-75. New York: St. Martin's, 1994.

———. *The Child Garden; or, a Low Comedy.* New York: Tor Books, 1989.

———. "A Fall of Angels: Or, on the Possibility of Life in Extreme Conditions." In *Unconquered Countries,* 1-100. New York: St. Martin's, 1994.

———. *Lust, or No Harm Done.* London: Flamingo, 2001.

———. "The Unconquered Country." In *Unconquered Countries,* 191-271. New York: St. Martin's, 1994.

Serres, Michel. "Redes." In *Atlas,* 139-49. Lisboa: Piaget, 1997.

Service, Robert F. "AFMs Wield Parts for Nanoconstruction." *Science* 282 (1998): 1620-21.

———. "Borrowing from Biology to Power the Petite." *Science* 283 (1999): 27-28.

Sheffield, Charles. "Comments: 'Deep Safari'." In *Nanodreams,* edited by Elton Elliott, 155-58. New York: Baen Books, 1995.

Simone, Rose. "UW Team Takes Big Steps in a Truly Tiny World: Nanotechnology Research Has the Potential to Shape Our Future." *The Record,* 24 October 2002. <http://therecord.com/business/techspot/business_techspot_02102315533.html>.

Site official du musée du Louvre. <http://www.louvre.fr/> (accessed May 25, 2003).

Smalley, Richard E. "Nanotech Growth." *Research and Development* 41, no. 7 (1999): 34-37.

———. "Of Chemistry, Love and Nanobots." *Scientific American* 285, no. 3 (2001): 76-77.

Smarr, Larry. "Nano Space: Microcosmos." *Wired* (June, 2003): 134.

Smithson, Robert. *Robert Smithson: The Collected Writings.* Edited by Jack Flam. Berkeley: University of California Press, 1996.

Snow, C.P. *The Two Cultures and the Scientific Revolution.* New York: Cambridge University Press, 1959.

Soja, Edward W. "Borders Unbound: Globalization, Regionalism, and the Postmetropolitan Transition." Forthcoming.

Sommerer, Crista, and Laurent Mignonneau, eds. *Art @ Science.* New York: SpringerWeinNewYork, 1998.

South West Museums Council (SWMC). <http://www.swmuseums.httpmedia.co.uk/> (accessed May 31, 2003).

Sparacino, Flavia. "The Museum Wearable: Real-Time Sensor-Driven Understanding of Visitors' Interests for Personalized Visually-Augmented Museum Experiences." In *Proceedings of Museums and the Web (MW 2002)*. Boston, April 17-20, 2002.

Stableford, Brian. "Great and Small." In *The Encyclopedia of Science Fiction*, edited by John Clute and Peter Nicholls, 518-20. New York: St. Martin's Press, 1993.

Stachel, John, ed. *Einstein's Miraculous Year: Five Papers That Changed the Face of Physics*. Princeton: Princeton University Press, 1998.

Stalder, Felix. "The Space of Flows: Notes on Emergence, Characteristics and Possible Impact on Physical Space." In *Proceedings of the 5th International PlaNet Congress*. Paris, August 26th - September 1st, 2001. <http://felix.openflows.org/html/space_of_flows.html> (accessed May 30, 2003).

Stephenson, Neal. *The Diamond Age: Or, a Young Lady's Illustrated Primer*. New York: Bantam Books, 1995. Reprint, New York: Spectra, 2000.

———. *Snow Crash*. New York: Bantam Books, 1992.

Sterling, Bruce. *Tomorrow Now: Envisioning the Next Fifty Years*. New York: Random House, 2002.

Stewart, Susan. *On Longing: Narratives of the Miniature, the Gigantic, the Souvenir, the Collection*. Baltimore: Johns Hopkins Univ Press, 1984.

Stix, Gary. "Little Big Science." *Scientific American* 285, no. 3 (2001): 32-37. Reprint in Fritz, comp., *Understanding Nanotechnology*, 6-16.

———. "Trends in Nanotechnology: Waiting for Breakthroughs." *Scientific American* 274, no. 4 (1996): 94-99.

Stonum, Gary Lee. *The Dickinson Sublime*. Madison: University of Wisconsin Press, 1990.

Strehle, Susan. *Fiction in the Quantum Universe*. Chapel Hill: University of North Carolina Press, 1992.

Sturgeon, Theodore. "Microcosmic God." In *The Science Fiction Hall of Fame: Volume I: The Greatest Science Fiction Stories of All Time*, edited by Robert Silverberg, 115-44. New York: Avon Books, 1971.

———. "Tandy's Story." In *The Norton Book of Science Fiction*, edited by Ursula K. Le Guin and Brian Attebery, 74-92. New York: W.W. Norton & Co., 1993.

Suvin, Darko. *Metamorphoses of Science Fiction: On the Poetics and History of a Literary Genre*. New Haven: Yale University Press, 1979.

Swanwick, Michael. "The Dog Said Bow-Wow." In *The Year's Best Science Fiction: Nineteenth Annual Collection*, edited by Gardner Dozois, 178-91. New York: St. Martin's Griffin, 2002.

Swiss Research Council. "Point of View: Transdisciplinarity on a Solid Basis." Annual Report, 1998, <http://www.snf.ch/en/por/phi/phi_view_hd1.asp>.

Tally, Steve. "A Shrinking World inside Agriculture." *Purdue Agricultures Magazine*, summer 2002. <http://www.agriculture.purdue.edu/agricultures/past/summer2002/features/feature_01_p2.html>.

Taylor, Richard. "Nanotech Time." *British Broadcasting Corporation*, 9 August 2003. <http://www.bbcworld.com/content/clickonline_archive_32_2001.asp?pageid=666&co_pageid=2>.

Thetis, Thomas N. "Letter to the Editor." *Technology Review* 102, no. 2 (1999): 15.

Timp, Gregory, ed. *Nanotechnology*. New York: Springer Verlag, 1999.

U.S. Congress. House of Representatives. Committee on Science. Subcommittee on Basic Research. *Nanotechnology: The State of Nanoscience and Its Prospects for the Next Decade: Hearing before the Subcommittee on Basic Research of the Committee on Science, House of Representatives, One Hundred Sixth Congress, First Session, June 22, 1999*. Washington, D.C.: U.S. Government Printing Office, 1999.

U.S. Congress. Senate. Committee on Commerce, Science and Transportation. Subcommittee on Science, Technology and Space. *New Technologies for a Sustainable World: Hearing before the Subcommittee on Science, Technology, and Space of the Committee on Commerce, Science, and Transportation, United States Senate, One Hundred Second Congress, Second Session, June 26, 1992*. Washington, D.C.: U.S. Government Printing Office, 1993.

Vesna, Victoria. Toward a Third Culture: Being in Between," *Leonardo*, 34.2 (April 2001). <http://vv.arts.ucla.edu/publications/publications_frameset.html>.

———. *Zero@wavefunction – nanodreams and nightmares*. <http://notime.arts.ucla.edu/zerowave> (accessed May 30, 2003).

Vinge, Vernon. "The Coming Technological Singularity: How to Survive in the Post-Human Era." Paper presented at the VISION-21 Symposium, NASA Lewis Research Center and the Ohio Aerospace Institute, March 30-31, 1993. <http://www.cse.ucsd.edu/users/goguen/misc/singularity.html> (accessed October 20, 2003).

Virilio, Paul. *The Art of the Motor*. Translated by Julie Rose. Minneapolis: University of Minnesota Press, 1995.

Virtual Museum of Canada. © CHIN 2003. <http://www.virtualmuseum.ca/English/index_flash.html> (accessed September 30, 2003).

Vorenberg, Sue. "Nanoscience: Big Interest in Studying the Very Small." *National Geographic News Online* (22 August 2002). <http://news.nationalgeographic.com/news/2002/08/0822_020822_nanoscience.html> (accessed October 15, 2003).

Voss, David. "Moses of the Nanoworld." *Technology Review* 102, no. 2 (1999): 60-62.

Walker Art Center. "New Media Initiatives." <http://www.walkerart.org/gallery9/> (accessed September 30, 2003).

Wang, Zhong Lin, Yi Lui, and Ze Zhang. *Handbook of Nanophase and Nanostructured Materials*. Vol. 2. New York: Kluwer Academic/Plenum Publishers, 2003.

Watson, Ian. *Nanoware Time*. (With John Varley. *The Persistence of Vision*). New York: Tor Double, 1991.

Webber, Andrew Lloyd and Tim Rice. "What's the Buzz/Strange Thing Mystifying," *Jesus Christ Superstar*. Decca, DXSA 7206, DL 79178, DL 79179 (3 discs). Sound recording, 1970.

Weiser, Mark, and John Seely Brown. "Designing Calm Technology." In *Xerox Parc*, December 21, 1995. <http://www.ubiq.com/weiser/calmtech/calmtech.htm> (accessed August 25, 2003).

Wertheim, Margaret. "Buckyballs and Screaming Cells: The Amazing Miniature World of Ucla Chemist Jim Gimzewski." *LA Weekly* April 4-10, 2003, 28-33.

Whitesides, George M., and J. Christopher Love. "The Art of Building Small." In Fritz, comp., *Understanding Nanotechnology*, 326-55.

Wiener, Norbert. *The Human Use of Human Beings: Cybernetics and Society*. Boston: Houghton Mifflin, 1954.

Williams, Raymond. *Culture and Society 1780-1950*. 1958. New York: Columbia University Press, 1983.

———. *Marxism and Literature*. New York: Oxford Press, 1985.

Windsor, Charles. "Acceptance Speech for the Grande Médaille. A Speech by the Prince of Wales." Société de Géographie, la Sorbonne, Paris, Thursday February 6, 2003.

Wolff, Cynthia Griffen. *Emily Dickinson*. Reading, MA: Addison Welsley, 1988.

World Transhumanist Association. <http://www.transhumanism.org/> (accessed 2003).

Wullschläger, Jackie. *Inventing Wonderland: The Lives and Fantasies of Lewis Carroll, Edward Lear, J. M. Barrie, Kenneth Grahame, and A. A. Milne*. New York: Free Press, 1995.

Zipes, Jack David. *The Trials & Tribulations of Little Red Riding Hood*. 2nd ed. New York: Routledge, 1993.

———. *Victorian Fairy Tales: The Revolt of the Fairies and Elves*. 2nd ed. New York: Methuen, 1987.

Zizek, Slavoj. "Bring Me My Philips Mental Jacket." *London Review of Books* 25, no. 10 (22 May 2003).

Zyvex Website. <http://zyvex.com/Products/home.html> (accessed October 20, 2003). <http://www.zyvex.com> (accessed 2003).

○○○

Contributors

Brian Attebery is the author of two award-winning books on fantasy literature and co-editor, with Ursula K. Le Guin, of *The Norton Book of Science Fiction.* His latest work is *Decoding Gender in Science Fiction,* published by Routledge in 2002.

Nathan Brown is a doctoral student in the UCLA English Department, with a BAH and MA from Queen's University, Canada. His primary fields of interest are contemporary poetry and poetics, critical theory and psychoanalysis, avant-garde modernisms, and science/technology studies.

N. Katherine Hayles, John Charles Hillis Professor of Literature in the English Department and Professor of Design/Media Arts at the University of California, Los Angeles, teaches and writes on relations of literature, science, and technology in the 20th and 21st centuries. Her recent book *How We Became Posthuman* (Chicago, 1999) won the Rene Wellek Prize for the Best Book in Literary Theory for 1998-99, and her latest book *Writing Machines* (MIT Press, 2002) won the Susanne Langer Award for Outstanding Scholarship.

Brooks Landon teaches in the University of Iowa English Department where he is Professor and Chair. His most book is *Science Fiction After 1900: From the Steam Man to the Stars* (New York: Routledge, 2002).

Susan E. Lewak is a doctoral student in English literature at the University of California, Los Angeles. She is currently focusing on the intersections of wonder, literature, and digital media from the golden age of children's literature to the present.

Kate Marshall is a graduate student in the UCLA English Department. Her research interests include twentieth-centuiry American literature, critical theory, and issues of technology, communication, and risk. She has written for technology business publications in England and the U.S.

Colin Milburn is a doctoral student at Harvard University in the departments of History of Science and English and American Literature and Language. His research focuses on the history of the biological and physical sciences, the Gothic novel, science fiction and posthumanism. He is currently writing a dissertation on the history of monsters in the modern era.

Jessica Pressman is a doctoral candidate in English at UCLA, where she is writing a dissertation about electronic literature. She is Associate Director of the Electronic Literature Organization www.eliterature.org.

Adriana de Souza e Silva is a Doctoral Candidate in Communications and Culture at the Federal University of Rio de Janeiro (UFRJ), Brazil. Since August 2001 she is a visiting scholar at de Department of Design | Media Arts at UCLA. Her research, "Hybrid Space Nomads", studies the impact of nomadic technology devices on our perception of spaces.

Carol Ann Wald is a doctoral candidate in the UCLA Department of English, where she specializes in science studies. Her dissertation focuses on evolutionary narratives in science fiction about robots and artificial intelligence. In 2001 she received the Bruns Prize for the best graduate student essay from the Society for Literature and Science.